1986

Agricultural Ecosystems

Agricultural Ecosystems
Unifying Concepts

Edited by
RICHARD LOWRANCE,
BENJAMIN R. STINNER,
and
GARFIELD J. HOUSE

A Wiley-Interscience Publication
JOHN WILEY & SONS
New York • Chichester • Brisbane • Toronto • Singapore

Library of Congress Cataloging in Publication Data:

Main entry under title:
Agricultural ecosystems.

"A Wiley interscience publication."
Includes index.
1. Agricultural ecology. I. Lowrance, Richard.
II. Stinner, Benjamin R. III. House, Garfield J.

S589.7.A36 1984   630.2′ 745   83-23504
ISBN 0-471-87888-X

Printed in the United States of America

10 9 8 7 6 5 4 3 2

Contributors

C. Vern Cole
  USDA-ARS
  Colorado State University
  Fort Collins, Colorado

David C. Coleman
  Natural Resource Ecology Lab
  Colorado State University
  Fort Collins, Colorado

George W. Cox
  Department of Biology
  San Diego State University
  San Diego, California

D. A. Crossley, Jr.
  Department of Entomology
  University of Georgia
  Athens, Georgia

Melvin I. Dyer
  Oak Ridge National
    Laboratories
  Oak Ridge, Tennessee

Edward T. Elliott
  Natural Resource Ecology Lab
  Colorado State University
  Fort Collins, Colorado

Robert D. Hart
  Winrock International
  Morrilton, Arizona

Garfield J. House
  Department of Entomology
  North Carolina State
    University
  Raleigh, North Carolina

Wes Jackson
  The Land Institute
  Salina, Kansas

John R. Krummel
  Oak Ridge National
    Laboratories
  Oak Ridge, Tennessee

George Langdale
  USDA-ARS
  Southern Piedmont
    Conservation Research
    Center
  Watkinsville, Georgia

Richard Lowrance
  USDA-ARS
  Southeast Watershed
    Research Lab
  Tifton, Georgia

Rodger Mitchell
  Department of Zoology
  Ohio State University
  Columbus, Ohio

Eugene P. Odum
  Institute of Ecology
  University of Georgia
  Athens, Georgia

David Pimentel
  Department of Entomology
  Cornell University
  Ithaca, New York

Edward J. Rykiel, Jr.
  Range Science Department
  Texas A&M University
  College Station, Texas

Renate M. Snider
  Department of Zoology
  Michigan State University
  East Lansing, Michigan

Richard J. Snider
  Department of Zoology
  Michigan State University
  East Lansing, Michigan

Colin R. W. Spedding
  Department of Agriculture
    and Horticulture
  Reading University
  Reading, United Kingdom

Benjamin R. Stinner
  Ohio Agricultural Research
    Development Center
  Wooster, Ohio

Robert Woodmansee
  Range Science Department
  Colorado State University
  Fort Collins, Colorado

Preface

The chapters in this volume were presented initially at a symposium on August 11, 1982, held during the Ecological Society of America meetings, State College, Pennsylvania. Three of the chapters (Crossley et al., Spedding, and Jackson) were incorporated later to complete the volume. The purpose of the symposium was to bring together expertise on diverse aspects of ecology as they relate to agriculture. It is intended that the meeting and the resulting papers help develop agroecosystem concepts, provide a foundation for future research, and, most important, encourage interaction between agricultural and ecological scientists.

Just as ecosystem research is generally carried out by a team of scientists, the organization of a day-long symposium and the publication of 13 chapters of new thoughts on agroecosystems is a team effort with all three editors doing an equal share of the work. We would like to express our thanks to the Ecological Society of America, and especially Dennis Knight, for sponsoring the symposium; to the Institute of Ecology, University of Georgia, for supporting the organization of the symposium; to the Southeast Watershed Research Lab, USDA-ARS, for support in editing manuscripts; and to the National Science Foundation for supporting agroecosystem studies in which the editors and contributors have participated. We hope that this collection of ideas on the unifying concepts of agroecosystems will be helpful to the students, teachers, and researchers working to increase the level of interaction between agricultural and ecological scientists.

RICHARD LOWRANCE
BENJAMIN R. STINNER
GARFIELD J. HOUSE

Tifton, Georgia
Wooster, Ohio
Raleigh, North Carolina
February 1984

Contents

INTRODUCTION   1

PROPERTIES OF AGROECOSYSTEMS   5
  Eugene P. Odum

THE ECOLOGICAL BASIS FOR COMPARATIVE PRIMARY
PRODUCTION   13
  Rodger Mitchell

CONSUMERS IN AGROECOSYSTEMS:
A LANDSCAPE PERSPECTIVE   55
  John R. Krummel and Melvin I. Dyer

THE POSITIVE INTERACTIONS IN AGROECOSYSTEMS   73
  D. A. Crossley, Jr., Garfield J. House, Renate M. Snider,
  Richard J. Snider, and Benjamin R. Stinner

DECOMPOSITION, ORGANIC MATTER TURNOVER, AND
NUTRIENT DYNAMICS IN AGROECOSYSTEMS   83
  David C. Coleman, C. Vern Cole, and Edward T. Elliott

AGROECOSYSTEM DETERMINANTS   105
  Robert D. Hart

ENERGY FLOW IN AGROECOSYSTEMS   121
  David Pimentel

EFFECTS OF SOIL EROSION ON AGROECOSYSTEMS
OF THE HUMID UNITED STATES   133
  George W. Langdale and Richard Lowrance

COMPARATIVE NUTRIENT CYCLES OF NATURAL AND
AGRICULTURAL ECOSYSTEMS: A STEP TOWARD PRINCIPLES   145
  Robert G. Woodmansee

MODELING AGROECOSYSTEMS: LESSONS FROM ECOLOGY   157
  Edward J. Rykiel, Jr.

AGRICULTURAL SYSTEMS AND THE ROLE OF MODELING   179
  Colin R. W. Spedding

THE LINKAGE OF INPUTS TO OUTPUTS IN AGROECOSYSTEMS   187
  George W. Cox

**TOWARD A UNIFYING CONCEPT FOR AN ECOLOGICAL
AGRICULTURE**   209
  Wes Jackson

INDEX   223

Agricultural Ecosystems

Introduction

The individual scientific contributions in this book examine the structure and function of agricultural ecosystems. Within the past decade agroecosystems research has emerged as an innovative and holistic science concerned with both basic and applied hypotheses. Ecologists are successfully addressing basic questions within the agroecosystem context, and agricultural scientists are finding ecological concepts valuable in solving management problems. Part of the impetus for increased use of the agroecosystem concept is the realization among some agricultural scientists that specialization within agriculture has led to reductionistic approaches which, in some instances, have created rather than solved problems. Also, agricultural research has been criticized recently as suffering from a lack of emphasis on basic research, where technological applications have surpassed fundamental information. On the other hand basic ecological research has not communicated and applied information to agriculture. Ecologists, as a group, have been reluctant to use agroecosystems as basic units of study due perhaps to the inherent complexity of systems controlled by both natural and social forces. It is our hope that the ideas presented in this book help reconcile these problems.

The major theme of this volume is the application of an ecosystem paradigm to agricultural science. While concise definitions for agroecosystems have been given, functional definitions often depend on the level of research or management of interest. Different levels can be determined by geographical, economic, agronomic, or ecological considerations. The array of topics presented here range from field-level agroecosystems (Crossley et al., Hart) to the farm-level (Cox and Spedding) to watershed agroecosystems (Woodmansee) to landscapes (Krummel and Dyer). This symposium does not address the question of the "correct" definition of agroecosystems since the research and management objectives determine the correctness. In fact, the authors disagree on a definition for an agroecosystem, some quite explicitly. Although a definition of agricultural ecosystem is relative and a function of research or management goals, an ecosystem approach implies that an agronomic unit is perceived as being comprised of interacting components that form a whole which has system-level properties. The logical extension is that to understand behavior of system components (crops, livestock, pests, etc.); one must know something of the way a component is connected within an ecosystem.

As Odum points out in his paper, agroecosystems are characterized by "extensive dependence and impact on externals; that is, they have both large input and output environments." Spedding extends this idea of external dependence to assert that agroecosystems "are essentially economic in nature." Whether agroecosystems are ecological systems under a high degree of socioeconomic control or agroecosystems are essentially socioeconomic systems with varying levels of ecological control is more than a semantic question. If agroecosystems are essentially socioeconomic systems as Spedding asserts, the study of ecological controls must be done within an existing or projected socioeconomic framework in order to be useful in the development of ecosystem management. Rykiel suggests that agroecosystems are actually "supersystems" that have "ecological, economic, and sociological components." Spedding sees an "interesting parallel" between the evolution of farms and organisms. This idea can be extended to consider the failure of farm agroecosystems as analogous to the extinction of organisms. In the past organisms and environments changed slowly together, allowing adaptations through evolutionary selection. Agricultural systems have always changed in response to technological innovation and changing social orders, but today's changes are different in both rate and kind for both organisms and agroecosystems. Rates are different due partly to society's increased emphasis on technological control of nature and partly to the general worldwide acceleration of technological change. The kinds of changes are different for both organisms and farms. Today's socioeconomic and technological changes tend to tie farms more closely to an economic network (e.g., fire-hardened digging sticks did not tie farmers to an uncontrollable economic system, 12-row no-till planters do).

From the researcher's perspective a fundamental difference between ecosystems and agroecosystems is that the former functions as a result of internal checks and balances. Thus, ecologists are able to create conceptual constructs, which they call "ecosystems." Agroecosystems, on the other hand, are not simply there; they are created in the physical sense by human intervention. Creation of agricultural ecosystems is necessarily concerned with economic goals—production, productivity, and conservation. Agroecosystems, by definition, are controlled by management of ecological processes. Jackson suggests a greater interaction between agronomy, population biology/ecology, and ecosystem ecology to bring about a sustainable agricultural complex. Our theme is that this sustainability can be accomplished through agroecosystem management—incorporating ecological, social, and economic goals into the design of sustainable agroecosystems for specific portions of the landscape. Therefore, agricultural management can evolve into agroecosystem management by broadening the set of goals which we seek to reach. These goals should be complementary to the traditional agricultural goals of productivity, production, and conservation.

We organized this volume on the framework of an ecosystem paradigm that consists of internal processes of primary production, consumption, and decomposition interacting with abiotic components and resulting in energy flow and nutrient cycling. Because of the anthropogenic character of agroecosystems, economic, social, and environmental determinants were added to this fundamental concept. This general model was chosen because it is flexible and should be applicable for both research and management purposes at different spatial scales (e.g., a corn field or a physiographic region).

After an even quick perusal of this volume, it should be obvious that the papers differ in basic character. Some of the contributions are data intensive while others are more conceptual. The authors express a diversity of perceptions that is expected given the diverse backgrounds of the participants and is appropriate due to the subject matter they are presenting.

Mitchell presents a comprehensive comparative approach to understanding primary production and other plant processes in different kinds of managed ecosystems. He establishes a definition of crop production parameters, so that meaningful comparisons can be made among differing kinds of production systems. Particularly interesting are the contrasts between intensive agriculture and village systems. He stresses that village systems employing mixed cropping methods are in many ways more stable and sustainable than monocultures.

Although the Woodmansee and Coleman et al. chapters both deal with nutrient relationships, the former focuses on overall ecosystem response to perturbation and various management practices while the latter explores interrelationships within decomposition processes. Both chapters point to reduced tillage as being more nutrient conservative because nutrient release from decomposition is more timed to uptake requirements of crops.

Pimentel explores the energetics of crop production by first presenting a historical perspective and then contrasting different types of cultural practices in terms of energetic efficiencies. He too cites advantages to reduced tillage practices, although the conclusions are based on energetics rather than nutrient cycling. Cox's paper also stresses efficiency of crop production but with less emphasis on energetics and more on economics. In addition, he provides useful discussion of crop ecological genetics and soil erosion all within the context of input–output behavior of agroecosystems.

The role of animals in agricultural systems is interestingly presented by Krummel and Dyer. Differing with traditional views of pests and animal production, they discuss management strategies in relation to the hierarchical nature of ecosystems, separating spatial and functional phenomena. The Crossley et al. chapter also argues a unique perspective on noncrop organisms. They emphasize a dearth of information on mutualistic or cooperative interactions that occur in managed ecosystems.

Ecosystem-level research and management entails understanding and manipulating complex interactions among physical, chemical, and biological processes where modeling and system-analysis tools are employed to tease apart cause-and-effect relationships. The papers by Spedding and Rykiel address this important aspect of agroecosystem science. Rykiel's approach is to discuss systems science and then apply it to the analysis of agricultural ecosystems. In contrast, Spedding's paper is oriented towards management and development of models for decision making.

Because of their anthropogenic nature, it is difficult to discuss agricultural systems and not comment upon human value systems. In reality, it is probably artificial to make such a separation. The first (Odum) and last (Jackson) chapters and Langdale and Lowrance's contribution on erosion explore issues that have very direct economic and political ramifications. All three authors converge on the concept of long-term stability or sustainability of crop production.

Properties
of Agroecosystems

Eugene P. Odum

Institute of Ecology
University of Georgia
Athens, Georgia

Agroecosystems are domesticated ecosystems that are in many basic ways intermediate between natural ecosystems, such as grasslands and forests on the one hand, and fabricated ecosystems, such as cities on the other hand. They are solar powered as are natural ecosystems, but differ in that (1) the auxiliary energy sources that enhance productivity are processed fuels (along with animal and human labor) rather than natural energies; (2) diversity is greatly reduced by human management in order to maximize yield of specific food in other products; (3) the dominant plants and animals are under artificial rather than natural selection; and (4) control is external and goal oriented rather than internal via subsystem feedback as in natural ecosystems (Fig. 1) (see also Patten and Odum, 1981, for a discussion of cybernetics of ecosystems).

Agroecosystems resemble urban-industrial systems in their extensive dependence and impact on externals; that is, they both have large input and output environments (Fig. 2). Agroecosystems differ in being autotrophic rather than heterotrophic. The power density level (rate of energy flow per unit area) of pre-industrial agriculture, as practiced in economically undeveloped countries, is not much different from that of natural ecosystems. Power density of industrialized agriculture is 10-fold or more greater than that of most natural ecosystems due to the high energy and chemical subsi-

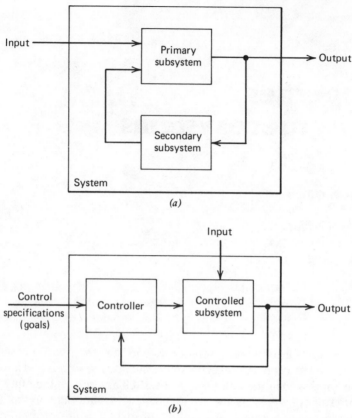

Figure 1. Natural ecosystems (a) are controlled by diffuse subsystem feedback in contrast to organisms and man-made systems (b) which have goals or set-points (after Patten and Odum, 1981).

dies. Accordingly, the impact of agricultural chemical pollutants and soil erosion on waterways, the atmosphere, and other global life-support systems can approach in severity that of urban-industrial areas.

Given the increasing cost of both energy and pollution, many agree that major technological, economic, and political efforts must be made to reduce the input and output costs of both agricultural and urban systems; otherwise, excesses in either or both will very soon jeopardize the capacity of the natural life-support systems to support them. Viewing croplands and pasturelands (and also plantation forestlands) as dependent ecosystems that are functional parts of larger regional and global ecosystems (i.e., a hierarchical approach) is the first step in bringing together the disciplines

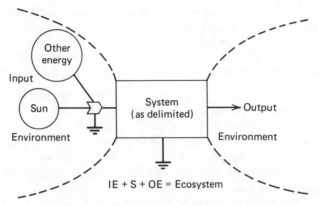

Figure 2. Input and output environments are a basic component of ecosystems which function as far-from-equilibrium systems. The properties of agroecosystems have changed dramatically as dependence and impact on externals has increased (after Odum, 1983).

necessary to accomplish long-term goals. The so-called world food problem cannot be mitigated by efforts of any one discipline, such as agronomy, working alone. Nor does ecology as a discipline offer any immediate or direct solutions, but the holistic and system-level approaches that underlie ecological theory can make a contribution to the integration of disciplines.

The properties of agroecosystems and the nature of their impact on other ecosystems have changed dramatically during the past half-century in the United States and other industrialized nations. It is important that we review this history in order to gain perspective on current problems and research needs. Auclair (1976) has described the development of intensive agriculture in the Midwest in three stages as follows:

1833–1934. Some 90% of prairie, 75% of wetlands and all forest land on good soils converted to croplands, pastures and wood lots. Natural vegetation restricted to steep land and shallow, infertile soils. However, farms were generally small, crops diversified, human and animal labor extensive, so impacts of farming on water, soil and air quality was not overall deleterious.

1934–1961. Intensification of farming associated with inexpensive fuel and chemical subsidies, mechanization, and increase in crop specialization and monoculture. Total cropland acreage decreased and forest cover increased 10% as more food was harvested from fewer acres by fewer farmers.

1961–1980. Increase in energy subsidy, size of farm, and farming intensity on the best soils, with emphasis on continuous culture of grain and soybean cash crops, much of which is grown for export trade. Conservation practices—such as crop rotation, fallowing, terracing, vegetated runoff waterways, etc.—decreased as farmers were more and more forced to expand cash crops to pay for increasing costs of energy and machinery. Yields per unit area increased but for some grain crops peaked during this period. Losses of farmland to urbanization and soil erosion accelerated, as also did the decline in water quality due to excessive fertilizer and pesticide runoff.

Brugam's (1978) analysis of the chemical composition of dated sections of cores from the bottom of a lake in Connecticut (Linsley Pond) provides a history of the changing impact of both agriculture and urbanization on adjacent ecosystems. Early farming in the 1800s had very little effect on the lake, but intensification of agriculture after about 1915 caused eutrophication of the lake resulting from an inflow of agricultural chemicals. From 1960 to the present, rapid urbanization and increased farming intensity has resulted in "hypereutrophication" due to agroindustrial wastes and extensive erosion that brought large amounts of soil, heavy metals, and other toxic substances into the lake. Marked changes in the biota directly related to changes in the input environment have been documented in this lake.

In summary, market and other economic and political forces, along with urbanization and the pressures of human population growth, have transformed agroecosystems from "domesticated" ecosystems that were relatively harmonious with our general environment into increasingly "fabricated" ecosystems that more and more resemble urban-industrial ecosystems when it comes to energy and material demands and waste production.

From the ecological perspective agroecosystems, coupled with natural ecosystems, constitute the human life-support module for spaceship earth since they provide the food, the water and air purification, and the other goods and services that sustain us. However, when agricultural products become valued more as market commodities to be sold to the highest bidder rather than as food to nourish us, and when short-term yields are maximized at the expense of long-term sustainable production, then the agroecosystem becomes more of a drain than a contribution to the life-support environment.

As the undesirability of some of these trends becomes evident, there is renewed interest in redeveloping conservation farming practices that first emerged in the 1930s following the soil erosion and dust bowl disasters. Among practices that should benefit from new technology and a better understanding of nutrient and water metabolism of crop systems are those that would (1) increase energy efficiency; (2) reduce irrigation water wastage and soil erosion; (3) increase nutrient retention and recycling so as to reduce fertilizer inputs; (4) promote use of crop residues for mulch, silage, and as energy sources; (5) increase diversity through multiple and rotational cropping; (6) reduce excessive dependence on broad-spectrum pesticides; and

(7) reduce plowing (limited-till and no-till). The latter alone can reduce costs of fuel and loss of soil by as much as 50% with only a moderate reduction of yield over the short term (Crosson, 1981). Theoretically, yields under no-till should exceed that under conventional till in the long-term due to reduction in erosion and improved maintenance of soil quality, but this theory is difficult to test when economics discourages investment in long-term experiments.

All of the conservation farming practices have the general effect of making the agroecosystem more like the natural ecosystem and less like the urban-industrial system, and hence a less disorderly and a more harmonious component of our total landscape.

On the basis of a national, interregional linear programming model, Olson et al. (1982) project that widespread adoption of what they term "organic farming" would increase net farm income and satisfy domestic demands for agricultural projects. Consumer food costs would increase, which benefits the farmer who would then have the means and motivation to maintain the quality of the land rather than "mine" the soil in a desperate effort to pay his debts. However, surpluses for export would decline sharply with widespread conservation farming. The dilemma here is that conservation farming is good for the farmer and the land, but not for the national economy of a nation committed (1) to exporting food to balance oil and mineral inputs and (2) to the manufacture of ever more farm machinery and agricultural chemicals.

Of the several characteristics of agroecosystems listed at the beginning of this chapter, I believe the one that merits special attention at this time is the manner of control (item 4). As we pointed out, control of agroecosystems is largely external, that of natural ecosystems to a considerable extent at least, internal. Subsystem controllers are more quickly responsive than external controllers to both internal feedback and external inputs. The independent, land-owning farmer, the backbone of American farming for most of our history, is an efficient "controller," as it were, since he is able to respond and adjust to local conditions and needs. His goal is not only to make a living but also to pass on his farm to the next generation in as good or better shape. To some extent, at least, such a farmer is an "internal controller" since he operates within the farming system. Unfortunately, in the past decade or so, control has more and more passed from the farmer to more distant controllers—absentee landlords, corporations, the federal government, and, especially, the grain and food markets. These remote controllers cannot respond effectively to the numerous positive and negative feedbacks that originate within the crop system itself. Furthermore, the goal of the remote controllers is primarily directed to obtaining the largest possible yield of a cash crop, not to maintaining long-term productivity.

In many ways, today's grain farmer, although much more affluent and better educated, is in the same frustrating position as the tenant farmer and sharecropper of yesterday's rural South who had to grow the same cash crop year after year even though he knew it was a vicious cycle that would soon impoverish both the land and himself.

Recognizing this basic problem is the first step in devising means to reestablish controls and goals to a more responsive local level in the hierarchy of agroecosystems. Beyond that, it may be feasible to design agroecosystems so that internal controls such as operate in natural ecosystems can contribute to overall efficiency, homeostasis, and stability. The theory here is that any services we can get from natural internal self-organizing and self-maintaining processes will reduce the need to spend money and energy to provide these services by artificial, external means.

Low-energy feedbacks that have high-energy effects are basic features of cybernetic systems. In ecosystems "downstream" components in the food chain such as predators or parasites may have a large-scale effect on primary production as a result of their control of herbivores, even though the predators and/or parasites utilize only a very small part of total community energy flow. Likewise, energy flow in a mycorrhizal network may be quantitatively small, but primary production of the whole ecosystem may be greatly enhanced by the direct soil to plant nutrient transfer work accomplished by the mycorrhizae. These are just two examples of potential subsystem controllers in natural ecosystems that can also operate in agroecosystems.

In our experimental research on agroecosystems at The University of Georgia, we are comparing ecosystem-level processes in conventional till and no-till cultivation of sorghum and soybeans with winter and spring rotations of rye or clover. Fertilizer applications are the same for both cultivations, and no insecticides or irrigation have been used on either. Just enough herbicides are used in no-till to prevent weeds from overtopping crops. For the first four years of our long-term experiment, crop yields have been very similar for the two treatments and comparable to yields obtained by farmers in the general region. In the no-till plots we are beginning to see some improvements in desirable properties such as nutrient and water retention and an increase in insect predators and parasites. We suspect, but have not yet demonstrated, that mycorrhizal networks developing in the upper layer of the undisturbed soil improve nutrient retention and may even link the root systems of volunteer plants and crop plants. If so, nutrients taken up by the former might become available to the latter (or vice versa). Accordingly, weeds between the rows or as an understory could be mutualistic rather than competitive with the crop.

In summary, viewing agroecosystems as intermediate between solar-powered natural ecosystems and fuel-powered urban-industrial systems helps to put current agricultural dilemmas in perspective. Increasing industrialization of agriculture has increased energy and chemical inputs, on the one hand, and chemical pollution and soil erosion outputs on the other hand. Control of the agroecosystem has become more remote (involving export market forces, absentee owners, and federal government), which results in maximizing short-term yield of cash crops at the expense of long-term production and maintenance of soil fertility. New forms of conservation tillage and a return to more local control are needed to reverse these undesirable trends. As a result of our own research and that of many others, we are encouraged to believe that reducing soil disturbance and toxic chemicals will allow natural mutualistic subsystems to develop that will improve the long-term fertility and stability of agroecosystems.

REFERENCES

Auclair, A. N. (1976). Ecological factors in the development of intensive management ecosystems in the midwestern United States. *Ecology* **57**:431–444.

Brugam, R. (1978). Human disturbance and the historical development of Linsley Pond. *Ecology* **59**:19–36.

Crosson, P. (1981). Conservation tillage and conventional tillage: a comparative assessment. Soil Conservation Society of America, Ankeny, Iowa.

Odum, E. P. (1983). *Basic Ecology*. Saunders Publishing Company, Philadelphia. pp. 16 and 79.

Olson, K. D., Langley, J., and Heady, E. O. (1982). Widespread adoption of organic farming practices: estimated impacts on U.S. agriculture. *J. Soil Water Conserv*. **37**:41–45.

Patten, B. C. and Odum, E. P. (1981). The cybernetic nature of ecosystems. *Am. Nat.* **118**:886–895.

The Ecological Basis for Comparative Primary Production

Rodger Mitchell

Department of Zoology
Ohio State University
Columbus, Ohio

Comparisons of primary production in nature and agriculture have done little more than list values for primary production without considering the significance of the differences or similarities (e.g., Westlake, 1963; Ovington et al., 1963; Cooper, 1975). There is now enough general information to go beyond simple comparisons and begin to consider what sorts of biological interactions might account for the observed levels of primary production in crops and natural communities. The production of natural communities stands as a useful basis for evaluating the biological performance of crop communities, but there are obvious technical problems in making comparisons because the basic field data for the production of the two systems are not directly comparable. The main body of agricultural data is a virtually complete record of the economic yields from genetically uniform mono-crops established on bare ground and maintained with extensive human intervention. Natural communities are freely interacting multispecies populations of mostly perennials for which ecologists have a scattered set of estimates of aboveground net production (ANP) and a much smaller set of data on belowground production. Hence, the first step in developing comparative primary production is to decide upon a representative measure of production that can be obtained from the ecological measures of ANP for natural communities and agricultural data on economic yields.

Once a practical basis for comparison is specified, the ecological factors that can affect yields must be outlined in order to begin to understand what might be inferred from observed differences in primary production. These problems can be resolved at a very general level, and the resulting comparisons appear to reveal overall patterns that suggest functional explanations for the broad differences in production. In some instances, comparative studies of the crop physiology of wild ancestors and derived cultivars reveal the biological basis for the evolution of yield in communities of cultivars (Eastin et al., 1969; Evans, 1975; Donald and Hamblin, 1976; Alvim and Kozlowski, 1977). These data on the evolution of cultivar phenotypes, competition, and community structure are of particular interest to community ecologists. My aim is to develop an ecological view of comparative primary production from a review of the current surveys of the agricultural literature on yields and summaries of the data on natural primary production.

AGROECOSYSTEMS

Agroecosystem research deals with a diverse range of systems and might be most properly defined as the use of the techniques and procedures of ecosystem analysis in the study of the organisms of an agricultural entity. The entity may be a single crop or a crop–consumer complex, such as grain and pigs. There are two kinds of agroecosystems, and these differ in the way production is linked with consumption and how that linkage, in turn, determines the distribution and abundances of cultivars. The distributions of crops and consumers in the large-scale commercial agriculture of developed countries are largely determined by intervening economic factors. The economic factors determining the abundances of crops and consumers in the nearly self-sustaining villages of less developed countries are the local demands for food, fodder, and fuel (i.e., a direct coupling of production with consumers). As a result of the direct coupling of production with consumption, village-level agriculture may be regulated by the villagers so as to function much like a natural ecosystem.

Village-level agriculture is quite unlike large-scale commercial agriculture in which the harvest is taken from the farm and enters the economic network which, in turn, determines the market from which consumers draw food and fiber. Thus, the economics of the marketplace, rather than biology, explains, for example, why American farmers plant wheat in areas where both the yield and the chances of a crop maturing are well below national averages (USDA, 1981). The linkage of production with the economic network is further strengthened when, as is commonly the case, the future yield is mortgaged to obtain the energy, chemicals, and machinery needed to manage the crop. Intensive monocropping cannot exist without an economic network; thus, it is the combined monocrop-economic network associations that are the functional determinants of the distribution, abundances, and, consequently, the realized yields of cultivars in developed countries.

The techniques of ecosystem analysis have been used in studies of the biotic interactions of the plants of a monocrop, and some authors define monocrops as agroecosystems (Loomis et al., 1971; Spedding, 1975; Loucks, 1977), even though monocrops are a dependent part of a functional crop-economic net regulated by economic functions that have no counterpart in natural ecosystems. The monocrops in commercial agriculture are certainly not special kinds of ecosystems, unless the definition of an ecosystem is narrowed so that it can include any monospecific population with its inputs and outputs.

The nearly self-sustaining villages that are still common in Asia, Africa, and South America are very much like the so-called natural systems ecologists treat as ecosystems. The territory of these villages is managed to produce the food, fodder, and fuel needed by the humans and other animals of the village as well as cash crops. The cropping decisions, which are strongly influenced by the needs of the organisms in the village, are executed with human and animal energy drawn from previous crops. Many of these villages could persist with little change if outside economic inputs were stopped. Insofar as the energy and materials needed to maintain the village as a discrete ecological entity are obtained from local biomass production, a village fits the explicit analytical ecosystem model developed from Lindeman's (1942) concept of trophicdynamics.

Just as dragonflies are neither dragons nor members of the order Diptera, agroecosystems are neither true ecosystems nor agricultural entities. In all agroecosystems economic values play some role in determining the distributions and abundances of cultivars. Economics is an overriding determinant of the distribution and abundances of cultivars in large-scale monocropping, whereas the direct biological linkages between producers and consumers play a direct role in determining the abundances of cultivars in village ecosystems.

PRIMARY PRODUCTION

Economic yields, aboveground net production (ANP), and net primary production (NPP) are part of gross production. Gross production is easy to define abstractly but probably impossible to measure with either experimental or field data. ANP will be used below because it is measured most often and more accurately than any other component of production. Gross production would be the best way to estimate the total biological activity of a plant community, but there is no way to obtain direct measures of gross production over long time periods. Still, the concept of gross production is an important concept because all measures of production can be defined in terms of the components in the energy budget for gross production.

Gross Production = Assimilation = Respiration + Growth

The natural time units for the gross production of a plant community is either the annual growth cycle or the life cycle of the plants. Physiological processes, assimilation and respiration, are measurable only over physiological time, minutes or fractions of seconds under uniform experimental conditions, and are appropriately expressed as rates of change over periods of minutes. It has proven to be impossible to correlate the rates for physiological processes obtained under controlled conditions with ecological measures of growth, which are the integrated outcome of photosynthesis, growth, and respiration acting and reacting in nature over time periods five orders of magnitude (yr = 5.2×10^5 min) greater than physiological time units (Monteith, 1977).

The energy budget may remain an elusive abstraction, but it is a useful abstraction for clarifying the differences between agricultural and ecological data and defining a reasonable basis for comparing production. ANP can be estimated from direct field samples and is not distorted by systematic errors of conversion (Table 1). ANP is a primary data base in all ecological studies. NPP is nearly always an extrapolation from ANP based on constants for the assumed relation between above and below ground production. Richardson (1978) and Sims and Coupland (1979) review sets of measurements of ANP and below ground production for marshes and grasslands, and these data do not show a significant correlation between the two components of growth. The extrapolation of root production is always based on relatively few samples and has undetermined systematic sampling errors that usually result in underestimates (Whittaker and Marks, 1975). Hence, the basis for extrapolating of NPP from ANP has not yet been demonstrated to be reliable.

ANP is the most common direct measure of production in natural communities, and some have argued that the use of ANP is justified on ecological grounds because ANP is ". . . a major component of energy that drives ecosystem processes . . ." (Webb et al., 1983). That may be true, but few would argue that there is sound evidence for relegating below ground pro-

Table 1.   Various approximations and practical measures of primary production in relation to the energy budget. Net primary production (NPP) is the accumulation of biomass, and aboveground net production (ANP) is the only accurately measurable biomass of terrestrial plants. Root production is generally estimated from constants (k_G) based on estimates of root/ANP ratios. Respiration and assimilation cannot be measured simultaneously over long time periods, hence, respiration is estimated with a constant (k_R) which is the ratio of respiration to biomass. The harvest index (h.i. = product/ANP) is usually used by agronomists in estimating ANP. When ANP is specified it implies that the other terms refer to aboveground components.

Measures of Biomass	Relation to Energy Budget
NPP	Growth = Assimilation − Respiration
Agricultural yields	
Commercial	Production = Assimilation − Respiration + (Growth − Product)
Extrapolation	ANP = Product + Product([1 − h.i.]/h.i.)
	= Assimilation − Respiration
Ecological yields	
Practical	ANP = Assimilation − (Respiration + Roots)
Extrapolated NPP	NPP = ANP + (ANP × k_G) = Assimilation − Respiration
Gross production	GPP = ANP + (ANP × k_G) + (ANP × k_R)

cesses to a minor role. At present, ANP is the only data set large enough to use in searching for patterns in productivity. The ANP of agricultural communities is rarely measured, but it can be extrapolated from the harvest index (h.i. = crop yield/ANP) of a cultivar. Because h.i. is directly measured for genetically uniform cultivars, it is likely to introduce only relatively small errors into the extrapolation of crop ANP from economic yields (see the excellent reviews by Loomis and Gerakis, 1975 and Donald and Hamblin, 1976). ANP is the best basis for comparisons because it is the usual data base in ecology and can be obtained from agricultural yields with reliable cultivar-specific constants.

The ANP of any community will vary with the environment. Five major environmental factors need to be considered in making comparisons of ANP: climate, nutrients, year-to-year variation in production, community structure, and time scale.

Climate

There are orderly patterns associating extrapolated ANP with climate. Ecologists have used a global data base to determine the way temperature and some measure of water availability correlates with production (Leith, 1975,

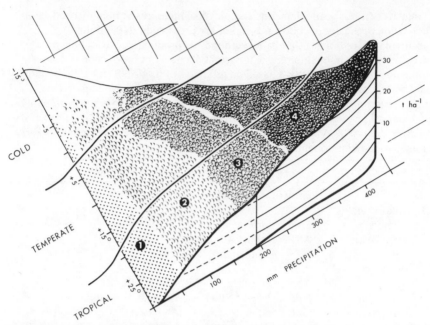

Figure 1. Global patterns of ANP as a function of temperature and precipitation. The approximate distribution of deserts (1), grasslands (2), savannahs (3), and forests (4) are indicated. Taken from a diagram of Whittaker's (1975) with the dimension of production added.

1976) and plant communities. These correlations account for a significant portion of the global variability in yield and are often projected on graphs to show how the distribution of plant communities correlates with climate. The dimension of NPP is added to Whittaker's (1975) graph (Fig. 1) to show how community type and production correlate with climate. Some correlations of the biotic responses of the communities can be added to these simple correlations of production with climate to obtain greater precision and, presumably, insight into the functions that determine production (Webb et al., 1983).

Relationships similar to these global correlations can be found on a local scale for individual cultivars. Thompson (1975) showed that the yield of wheat correlated with the average temperature and rainfall (Fig. 2). The Kansas wheat harvest is nearly twice as large as any other state, despite yields that are far below the potential yield of wheat. Recent yields for U.S. wheat production, 2.1 tonnes ha^{-1}, are far below those of Europe, 3.5 tonnes ha^{-1}, because economic factors exclude wheat from the climates in which its potential yield could be realized. Such economic intervention alters agri-

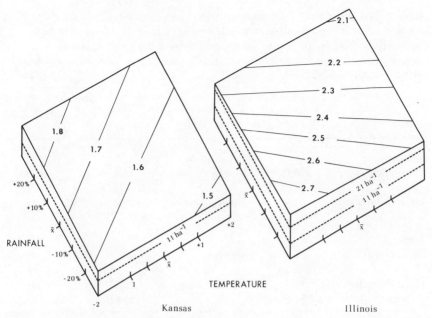

Figure 2.   The economic yields of wheat in response to variations in weather in Kansas and Illinois as predicted from the models of Thompson (1975).

cultural production to an extent that agricultural statistics do not indicate either the yield potential or the relative potential of the farming practices of an area. It is appropriate to use the optimal crop production values, such as given in the careful review of Loomis and Gerakis (1975), in comparisons that are meant to measure the effect of biological differences between crops and natural communities.

Nutrients

Averages for production by community and climate (Figs. 1 and 2) usually reveal orderly responses on a global scale. However, such broad-scale correlations obscure the systematic deviations produced by the unsubsidized and naturally subsidized nutrient cycling regimens that Odum (1975) distinguished. A few habitats (flood plains, marshes, estuaries, upwelling areas of lakes and oceans, and swamps) receive natural subsidies of nutrients that usually produce a substantial increase in primary production (Table 2). Agricultural production is sometimes compared to adjacent unsubsidized natural ecosystems (e.g., Ovington et al., 1963), but this reveals more about the role of nutrient subsidies than the differences between agricultural and natu-

Table 2. Aboveground net production (ANP) for plant communities without nutrient subsidies and those with natural nutrient subsidies. The growth rate is obtained by dividing the annual yield by the days of the growing season. The ranges are for what is often called the "usual" values, which means that isolated extreme values are ignored.

	Unsubsidized ANP ANP (tonne ha^{-1} yr^{-1})		Subsidized ANP ANP (tonne ha^{-1} yr^{-1})	Growth Rate (g m^{-2} yr^{-1})
Temperate				
Grasslands[a]	8.0 (2.2–14.2)	Marshes[b]	14.4 (1.5–21.0)	6.53 (2.5–13.9)
Upland forests[c]	11.2 (6–18)	Flood plains[d]	9.0 (4.8–17.8)	4.8 (2.8–7.9)
Tropical				
Grasslands[e]	8.1 (0.8–34)	Marshes	60[f] (20–100)	16.44 (5.5–27.4)
Upland forests[g]	11.1 (3.2–19.3)	Flood plains	29.3[h] (15–42)	8.0 (4.1–11.5)

[a] Sims and Coupland, 1979.
[b] Richardson, 1978; Brinson et al. 1981.
[c] Kira, 1975; Cannell, 1982.
[d] Brinson et al., 1981; with growing seasons from Lugo and Brinson, 1978.
[e] Murphy, 1975; Coupland, 1979.
[f] Midrange estimate based on Gaudet, 1977; Singh et al., 1979.
[g] Kira, 1975; Murphy, 1975.
[h] Rodin et al., 1975, assuming ANP as 0.6 of NPP.

ral communities. Nearly all agricultural production is nutrient subsidized with chemicals or manure and, consequently, should be compared to other naturally subsidized plant communities.

Annual Variation

There is a virtually complete record for year-to-year variation in agricultural yields, but there is very little information on the annual variation in natural ecosystems. Under the U.S. International Biological program one sequence of four year's data for forest productivity (Webb et al., 1983), one sequence of four year's data for forest productivity, and seven 3-year (1970–1972) sequences of primary production of American grasslands (Sims and Coupland, 1979) were obtained. The grassland data raise two disturbing points. The first, the failure of root production to correlate with ANP, has already been mentioned. Second, the differences between the highest and lowest ANP, as a fraction of the average for the three-year sequences, averages 47%. Agricultural data for hay and alfalfa production in Colorado (USDA, 1973–1974) during the same period show a maximum difference of 4%.

A rough indication of the possible association of variability in ANP relative to grain yields can be obtained by comparing the yields of maize silage with grain yields. Silage is an approximation of ANP because it is harvested after anthesis at which time vegetative growth is nearly complete and the photosynthetic capacity of the vegetative components, which will determine the potential grain yield (Duncan, 1975), is established. The yields of maize and silage in the 10 largest maize-producing states from 1975 through 1979 (USDA, 1977–1981) are closely correlated (r_{48} = 0.929): maize = 0.876 + (0.3429 × silage). The maximum difference in maize plus silage yield for the 10 three-year sequences was well below 50% in all but three sequences. The average maximum difference between yields for three-year periods was 36%.

At first glance, agricultural production data appear to be far less variable than natural communities, but there are not enough data for an evaluation of variability.

Community Structure

The commercial agricultural systems of developed countries are monocultures, whereas much of the dryland farming in village agriculture in Africa, Asia, and South America makes extensive use of the techniques of mixed cropping (Buck, 1930; Miracle, 1967; Grigg, 1974; Stelly, 1976; Norman, 1979; Webster and Wilson, 1980). Where rains are heavier, exceptional systems of 10–40 cultivars are grown together. These complex systems have not been extensively studied aside from yields, which are almost always larger or more reliable, or both, than those of monocrops (ICRISAT, 1981).

Natural communities are usually a diverse interacting complex of species, but the diversity of species is often sharply reduced in communities enjoying natural nutrient subsidies, such as marshes and swamps. Some marshes, for example *Typha* and *Spartina*, are nearly monospecific systems (McNaughton, 1966; Pomeroy et al., 1981) with a diversity and structure similar to that of agricultural systems.

The area of leaves over a given area of ground is designated as the leaf area index (LAI) and is a useful basis for comparing the photosynthetic area at a community level. It may be a crude index of the growth going into leaves and, as a rule, the greater the LAI the more competition there is for light.

Time Scale

Annual production is the parameter used in most ecosystem analyses, but the average crop growth rate, obtained by dividing the total yield by the days of the growing season, should be used in evaluating the biological performance of communities independent of the growing season length. The growth rates of naturally subsidized plant communities overlap more than ANP (Table 2). If one is concerned about the relation of growth rates to ANP, it is necessary to measure the variation in the rates of production over the growing season. When estimates of growth are made every few days or once a week, the maximum short-term growth rate achieved by a community can be distinguished. These measures of growth rates should not be confused with the rate of carbon fixation for isolated leaves.

The average growth rate, which is smaller than the maximum rate, may reflect seasonal interactions. The average growth rate for temperate communities may be less than that of tropical communities (Table 2) because it takes a part of the growing season for temperate communities to grow an efficient canopy for light absorption. Possibly the communities of annual cultivars would have a lower average growth rate than that of communities of perennials, which can grow a canopy quickly from the stores of photosynthate in roots. The effect of such ontogenetic changes on ANP can be evaluated by comparing the average growth rate with the maximum short-term growth rate.

Maximum rates are the complex outcome of interactions at three levels: molecular processes, leaf anatomy, and structural interactions that determine the geometry of the canopy of a community. Several analyses of these data (Sheehy and Cooper, 1973; Gifford, 1974; Monteith, 1977) show higher level interactions to be so important as to obscure the differences existing in the primary (molecular and chloroplast) processes of photosynthesis. The maxima fall in a narrow range. Average rates are well below the maxima and span a range as wide as the values for annual production (Table 3).

Table 3. Production (ANP) of crops ranging from legumes to tropical forage grasses.[a]

	Growth Rate (g m^{-2} day^{-1})		Crop ANP (tonne ha^{-1})
	Maximum	Average	
Crops	20–55[b]	0.5–24.5	2–68

[a] Based on Evans (1975), Loomis and Gerakis (1975), and Cooper (1975).
[b] Rates above 55 for carrot and sunflower have been questioned by Loomis and Gerakis (1975) on the basis of experimental procedures and the improbability of the implied solar energy conversion of 7.5%.

COMPARATIVE PRIMARY PRODUCTION

The aim of these comparisons is to gain insight into the responses of plant *communities,* which has been a goal of ecologists but has not been an important aspect of agronomic research until recently. Current books on crop physiology (Evans, 1975; Alvim and Kozlowski, 1977) supplemented by reviews of production and crop biology (Eastin et al., 1969; Loomis et al., 1971; Cooper, 1975; Loomis and Gerakis, 1975; Donald and Hamblin, 1976; Cannell, 1982) provide the central core of information used here. The values for agricultural production used in the following comparisons will be the yields currently attained under favorable conditions. The record yields and yields under favorable conditions are cited to indicate the limits attained by field crops, not the yields achieved in small experimental plots. When available, historic yields for agriculture in traditional village ecosystems will be given to indicate the yields attained under organic agriculture.

The production figures for natural systems are taken from reviews and summaries that cite sources and specify how average or typical values were obtained.

Rice and Marshes

Rice is a domesticated marsh plant that has displaced natural marshes over large areas from the tropics to temperate regions, so it is entirely appropriate to compare the productivity of natural marshes with rice paddies. The most thoroughly studied marsh communities (Table 4) are close to being monospecific stands of cloning perennials such as *Typha* and *Spartina*. The exception, *Zizania*, is an annual and may be the most abundant species in marshes with a high diversity.

The growing season for marshes ranges from 100 to 365 days, but average crop growth rates appear to be reasonably consistent (Table 4) with the majority of growth rates from virtually all communities in the range of 10–20 g m^{-2} day^{-1}. The significance of the single *Zizaniopsis* record cannot be judged, and the ANP for *C. papyrus* may be an inflated extrapolation from Gaudet's (1977) data on the production of a single cohort. However, papyrus ANP lies at the extreme for marshes as would be expected for a tropical marsh.

Clearly, contemporary rice and marsh plants have similar rates of productivity, and historical records indicate that rice ANP has not changed significantly except in Japan (Table 5). There has been an increase in economic yields as indicated by the h.i., which has increased from about 0.3 to the current values of 0.5 for high-yielding varieties. Most of that change occurred in the last 40 years as a result of plant breeders shifting from the techniques of mass selection to the a *priori* design of plant ideotypes (Uchijima,

Table 4. Productivity of rice and natural marshes. Rice data are for single crops, and the maximum crop growth rates are for the dominant species while the average and annual ANP are for natural marshes, which will have a mixture of species. The photosynthetic pathway (C_3 or C_4) is given when known.

	Growth Rate $(g\ m^{-2}\ day^{-1})$		ANP $(tonne\ ha^{-1}\ yr^{-1})$	
	Maximum	Average		Reference
Annuals				
Rice (C_3)	55	7–17	8–12	Evans, 1975
Record for rice	—	21.6[a]	26	Yoshida, 1977
Zizania aquatica	—	10.5	15.8	Good and Good, 1975
Perennials				
Cyperus papyrus (C_3)	41	12–19	40–60	Westlake, 1975 Gaudet, 1977
Spartina alterniflora (C_4)	64	2.2–15.5	5.0–40	Lugo and Brinson, 1978; Giurgevich, and Dunn, 1982
Typha latifolia (C_3)	53	2.0–12.5	3.3–19	Brinson et al., 1981
Zizaniopsis maliacea		6.2	15.3	Birch and Cooley, 1982
Temperate marsh ave.	10–53	2.5–13.9	1.5–21	Richardson, 1978; Brinson et al., 1981
Emergent C_3 marshes	10–13	12–48	40–60	Westlake, 1975

[a] Murata and Matsushima (1975) give 36 g as the theoretical maximum.

1976), which are obtained from crosses and tested under group selection. Artificial selection based on yield criteria alone is rather like an intensified natural selection in which there is selection for survival (successful competitors) and high fecundity among the survivors in segregating populations. Such selection can result in a slow improvement in yield but, at times, positive selection for yield can give a negative response. Such a negative response to selection for yield was demonstrated by Wiebe et al. (1963). Griffing (1967) has shown that the negative results are the result of an unbalanced selection, and Donald and Hamblin (1976) provide an incisive list of traits favored in competing populations; high crop yield is not one of those traits. Individual selection in segregating populations favors competitive high-yielding individuals, while group selection is based on the performance of a group of related individuals. Selection for yield can be carried out independently of competition by using group selection in which the se-

Table 5. Historical records of average rice production in favorable areas under animal-powered village agriculture without the use of chemicals. ANP and growth rates were based on assumed values of 0.30 for h.i. and a growing season of 150 days.

Source	Locality and Date	Rice (tonne ha^{-1})	ANP (tonne ha^{-1})	Growth Rate (g m^{-2} day^{-1})
Hsu, 1980	China, 87 B.C.–200 A.D.	3.47	11.6	7.7
Rawski, 1972	China, 1600–1900	2.1–3.4	7–11.3	6.1
Buck, 1930	China, 1926–1928	2.74	9.1	6.1
Ishizuka, 1971	Japan, 800–900	1.01	3.4	2.3
	1720	1.92	6.4	4.3
	1900–1917	2.64	8.8	5.9
Stanhill, 1981	Egypt, 1800	2.07	6.9	4.6

lection criteria are the performances of stands of genetically uniform individuals.

The development of the theoretical basis for the engineering of efficient non-competitive ideotypes, which started with the analyses of Monsi and Saeki (1953), has led to the quantum jump in yield achieved with high-yielding varieties. These ideotypes, which perform very much as predicted by theory (Uchijima, 1976), could not be obtained under the traditional pattern of selecting high-yielding individuals from competing populations (Donald and Hamblin, 1976).

It stands to reason that the traditional varieties produced by individual selection in competitive environments might have canopies and growth patterns similar to natural vegetation, while an engineered ideotype proven by group selection would have different traits. Some of the combinations of traits expected under various programs of selection (Donald and Hamblin, 1976) can be evaluated for rice, *Zizania*, and *Spartina*.

The interactions between rice ideotypes were examined in a set of experiments set up by Jennings and colleagues (1968). A variety of *indica* rice, designated BJ, was used to represent a competitive phenotype produced through individual selection. Pure stands of BJ develop a canopy that attains its maximum rate of carbon fixation about 60 days after planting (Fig. 3) and is followed by a decline in the rate of growth from 75 to 109 days after planting. During this decline the lower leaves become shaded as the LAI rises to 6.2. A similar reduction in growth rate is seen in natural marsh canopies dominated by perennial (*Typha*) and annual (*Zizania*) plants. *Typha* increases its growth rate rapidly and maintains its maximum rate, 20–22 g m^{-2} day^{-1}, for nearly 50 days before the growth rate begins to fall at about 75 days (Fig. 3). The persistence of the period of high growth rates may well reflect the

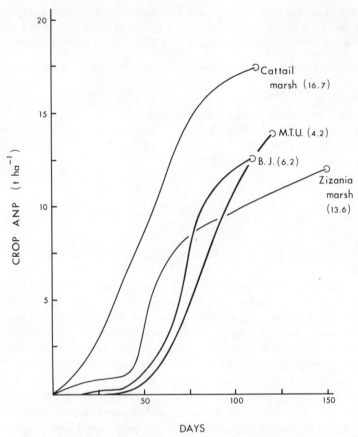

Figure 3. The seasonal pattern of growth for two varieties of rice and two natural marshes that approach being monospecific stands. BJ is a competitive tall variety of rice and MTU is a dwarf high-yielding variety. Based on data from Jennings and colleagues (1968), Jervis (1969), and Good and Good (1975).

advantages of a canopy of erect leaves that can increase rates of photosynthesis up to LAI values of 10 (Ishizuka, 1969; Monteith, 1969). A sigmoid growth curve is clearly exhibited by *Typha*, *Zizania*, and the competitive rice, BJ, which have large LAI values and rates of growth that fall below 10 g m^{-2} day^{-1} for about the last quarter of the season. The seasonal decline in the growth rate of *Spartina* appears to result from the erect leaves bending toward a horizontal position later in the season (Turitzen and Drake, 1981).

The non-competitive rice used by Jennings and colleagues (1968), MTU, attains the same maximum growth rate as competitive varieties with a lower LAI. A crop growth rate of over 20 g m^{-2} day is attained by day 60 and held until day 110. It still averages 15 g m^{-2} day^{-1} during the last 10 days of the season.

The maximum growth rates are similar in these four marsh communities (Fig. 3), but ANP differs as a function of how long maximum crop growth rates can be maintained. The high-yielding rice variety, MTU, developed by modified group selection is fundamentally different from that of the competitive rice variety and the natural species presumed to have evolved through individual selection (Jennings and Aquino, 1968) because the LAI is low and high rates of growth persist almost to the end of the crop cycle in high-yielding varieties.

The average crop growth rates are similar in natural marsh plants and rice. Maximum short-term rates do not exceed 55–65 g m^{-2} day^{-1}. The increases in the yields of crop plants have not resulted from an increased growth rate. Growth rates have remained the same, and the increased economic yield has resulted from reducing the photosynthate going into an unnecessarily dense canopy so that a larger portion can be diverted to the seeds. Economic yields increased while ANP has not changed and still remains at about the same levels as those of comparable natural communities.

Forage Crops

These communities resemble marshes, although they are, as a rule, upland crops. When these crops are heavily subsidized and grow throughout the year, their ANP reaches very high levels. Yields for subsidized forage grasses (Table 6) are similar to seasonal marshes, and napiergrass ANP is certainly comparable to that of papyrus (Table 4). The highest average commercial yields of hay in the United States are from Arizona (USDA, 1980) where most of the production is at least water subsidized, but even these yields are not much greater than those of the moist Grasslands Biome sites (Sims and Coupland, 1979). Hay production in the United States averages at about twice the natural ANP of the northern Biome sites (Table 6).

Gifford (1974) used forage crops to show that the differences in the photosynthetic rates of C$_3$ and C$_4$ plants at the biochemical level are obscured by interactions at the microscopic to the macroscopic level so that there is a broad overlap in the production of C$_3$ and C$_4$ plants. The rates of 20–40 g m^{-2} day^{-1} (Sheehy and Cooper, 1973) are for stands in small boxes under optimal greenhouse conditions.

Upland Seed Crops

These occupy areas that do not usually have natural nutrient subsidies. Cultivation inevitably alters the natural nutrient equilibrium of upland soils be-

Table 6. ANP of forage grasses under optimal conditions.

	Growth Rate (g m^{-2} day^{-1})		ANP (tonne ha^{-1} yr^{-1})
	Maximum	Average	
Bermudagrass (Cynodon dactylon C$_4$)[a]	—	7.5	23–32
Napiergrass (Pennisetum purpureum C$_4$)[a]	54[b]	23.4	85
Ryegrass (Lolium perenne C$_3$)[c]	20[c]	7.6	27–29
Alfalfa (Medicago sp C$_3$)[a]	—	7.8	28–29
Hay (Arizona)[d]		1.6	5.8
Lawns[e]			17
C$_3$ forage grasses[f]	16–43	6.0–13.2	21–33
C$_4$ forage grasses[f]	39–52	7.4–23.4	25–85
Grassland biome[f]			
Northern sites		2.8	2.45
Southern sites	—	0.9	0.20

[a] Loomis and Gerakis, 1975.
[b] Loomis et al., 1971.
[c] Gifford, 1974.
[d] USDA, 1980.
[e] Falk, 1980.
[f] Cooper, 1975.

cause of the loss of nutrients in the increased flow of water through disturbed soil. The nutrients were traditionally supplemented with manure, and the yields of this organic agriculture are distinguished from those of chemical agriculture (Table 7).

Village ecosystems have low productivity, and some assert that the low productivity is due to both a lack of chemical fertilizer and the inefficiency of human and animal power. Mechanization does not necessarily increase yields over those achieved under animal power (Singh and Chancellor, 1974, 1975), and in the United States mechanized organic farming produces corn and soybean yields that are only marginally lower than yields with chemical agriculture. Wheat is 43% lower (Lockeretz et al., 1981). Yields from intensely mechanized Japanese rice production are below those of the nearly completely animal-powered rice production in Korea (Lee, 1979). High-yielding varieties and chemicals are used in both countries, and so the difference would seem to be attributable to the greater care taken in animal agriculture. Those with extensive knowledge of animal-powered village agroecosystems in Asia and Africa assert that village ecosystems consistently achieve higher yields than large mechanized farms of the area when nutrient inputs and other biotic factors are similar (Moerman, 1968; Chandler, 1979; Wortman, 1980). Hence, records of production in village agroecosys-

Table 7. Productivity of upland seed crops. ANP values represent the usual maxima for contemporary agriculture based on Loomis and Gerakis (1975), Cooper (1975), and U.S. Department of Agriculture (1980). The organic ANP is taken from historical records (Buck, 1930; Lennard, 1932; Bath, 1963; Elston et al., 1980; Stanhill, 1981) and current village-scale ecosystems using neither chemical fertilizers or tractors (Kowal and Kassam, 1978; Norman, 1979). The growth rates (Evans, 1975) and the h.i. (Zelitch, 1975; Donald and Hamblin, 1976) are for chemical agriculture.

	Growth Rate $(g\ m^{-2}\ day^{-1})$		ANP $(tonne\ ha^{-1})$ Inputs		Harvest Index
	Maximum	Average	Chemical	Organic	
Barley C_3	—	11.8	6–18	2.5	0.55
Maize C_4	52	18–23	20–30	8.3	0.43
Millet C_4 (*P. typhoides*)	54	19.6	22	5.7	0.21
Oats C_3	—	7–15	3–6	4.9	0.41
Rye C_3	—	—	6–13	3.0	0.30
Sorghum C_4	51	12–18	10–15	3.6	0.41
Soybean C_3	27	6.7	8	3.1	0.32
Sunflower C_3	68[a]	6.7	10	—	0.57
Wheat C_3	22[b]	8–18	9–13	4.2	0.45

[a] Loomis and Gerakis (1975) give a value of 31 and question the experiments on which the higher values were based.
[b] Austin (1982).

tems do seem to give a fair indication of the potential yields achievable with only biotic inputs.

Yields for organic agriculture (Table 7) generally lie between 3 and 5 tonnes ha^{-1}, which is well below the usual level for unsubsidized natural communities (Tables 2 and 6). These estimates combine the more readily available records from Europe where crop rotation and manuring were rigorously employed only in the last 200 years with records from areas such as China where crop rotation and manuring were urged on the farmers by the government 2000 years ago (Hsu, 1980). When chemicals are used to raise the nutrient equilibrium, the ANP of all but oats increase to levels that are equal to or greater than those of nutrient unsubsidized natural and agricultural communities (Tables 2 and 7). In many crops, yields even exceed the average for naturally nutrient subsidized temperate communities. Historic yields of rice under organic agriculture are generally from naturally nutrient subsidized areas, and their ANP is above that of upland crops (Tables 4, 5, and 7).

The average growth rates of the majority of the plant communities fall into the fairly narrow range of 10–20 g m^{-2} day^{-1} quite independently of photosynthetic system, nutrient subsidies, or community type (Tables 4–7). Just as in rice, it appears that the potential ANP of upland seed crops has not changed, and the increases in economic yields are the result of high growth rates persisting longer and a larger portion of photosynthate going into grain rather than an increase in ANP. The history of the evolution of wheat yields and the way growth patterns have changed during this century is particularly clear, thanks to the comparative studies of Austin et al. (1980) who grew stands of the major winter wheat varieties released in Britain over the last 80 years under the same conditions. The community characteristics of the

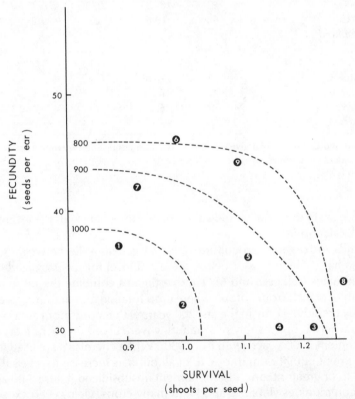

Figure 4. Components of Darwinian fitness of the major wheat cultivars released in England. The dotted lines indicate the density of shoots per m^2 in the mature stands. Numbers refer to varieties in sequence of their release: (1) Little Joss—1908; (2) Holdfast—1935; (3) Cappel-Desprenz—1953; (4) Maris-Widgeon—1964; (5) Maris-Huntsman—1972; (6) Hobbit—1977; (7) Mardler—1978; (8) Armada—1978; (9) Norman—1982. Data from Austin et al., 1980.

stands were carefully monitored and these data can be used to compare biological fitness, the product of survival and fecundity. The number of seeds planted is taken as the number of individuals beginning the generation, and the number of fertile shoots per seed is the survival. Survival can be greater than one because tillering is the equivalent of a phase of asexual reproduction between sexual generations. Fecundity is the number of seeds per ear (Fig. 4). Artificial selection for yield has acted on either survival or fecundity but not consistently on both. The only fairly consistent trend has been the lowering of shoot density, as the isopleths of Figure 4 show.

Yield is a complex joint function, as can be seen from a comparison of the traits exhibited by the first and last wheats in the series (Table 8). The ANP of the two wheats is virtually the same, but the economic yield of the most recent variety is much greater because of a 42% increase in h.i. The LAI is lower in the high-yielding ideotypes, and a reduction in plant density results in a more efficient canopy just as in the case of rice (Fig. 3). The criteria for designing high-yielding wheat ideotypes (Donald, 1968) has been much the same as that used for rice.

Perhaps the most astonishing finding in crop physiology is the fact that photosynthetic rate of wheat leaves has fallen during the evolution of wheat cultivars from their presumed ancestors (Evans and Dunstone, 1970; Dunstone et al., 1973). It is now well established that the wild relatives of wheat, pearl millet, sorghum, and cotton all have higher maximum light-saturated CO_2 exchange per leaf area than modern cultivars (Gifford and Evans, 1981). Thus, it appears that selection for increased crop growth rate, or yield, or both often results in a reduction in the rate of leaf photosynthesis. This phenomena remains unexplained and may be one more example of an apparently negative response to positive individual selection for yield (Wiebe et al., 1963; Griffing, 1967). Since several crops are involved, the reduction may reflect some as yet undetermined general interaction in the complex set of factors affecting photosynthetic processes and which is affected by artificial selection.

Table 8. Characteristics of British wheats. A variety released at the start of the century is compared to the most recently named variety using the data of Austin et al. (1980). Crop growth rates ranged from 12 to 18 g m^{-2} day^{-1} with no significant differences between varieties.

			tonnes ha^{-1}		
	ANP	Grain	Shoots	LAI	Fertile Shoots
Little Joss (1908)	14.5	5.22	1114	6.90	32%
Norman (1982)	15.0	7.57	813	5.88	51%

Upland Temperate Forests

Cannel's (1982) meticulous review of data for forest biomass shows a range of 3–32 tonnes ha^{-1} yr^{-1} (Table 9). The vast majority of the values fall from 8 to 13 tonnes ha^{-1} yr^{-1}. Except for the extremes, this is not greatly different from marshes (Table 4). Forests have been replaced by grains in much of the northern hemisphere, and the grains have about half the ANP under organic farming and can match the forests when chemical subsidies are given (Table 7). The ANP of tree plantations is not much different from that of natural forests. The range for maximum ANP, 20–30 tonnes ha^{-1} yr^{-1}, does not differ from the plant communities reviewed above. For a growing season of 160 days, this is an average crop growth rate of 12–18 g m^{-2} day^{-1}.

Long-Term Tropical Crops

These appear to have very high yields (Table 10), but the data are for crops with growing seasons of a year or more. Regrettably, there are little data for comparisons, but the growth rates do not depart appreciably from those of temperate crops.

Mixed Cropping

These systems cannot be evaluated in the same way as the monocrops that replace natural ecosystems on a regional basis, such as the corn replacing

Table 9. The ANP of upland temperate forests and tree plantations taken from Cannell (1982). The natural and unmanaged stands, usually second growth, are designated by the dominant species. There are inconsistencies because the original data vary with respect to the inclusion of minor species and litter fall. Plantation ANP generally excludes minor species and litter fall. When a set of samples were reported, the maximum value was taken for the tabulation.

	Natural/Unmanaged (tonne ha^{-1} yr^{-1})	Plantations (tonne ha^{-1} yr^{-1})
Conifers	3.0–32.2	8.1–24.6
Abies	10.1–26.8	8.7–18.6
Picea	3.2–11.0	7.5–14.5
Pinus	4.7–17.6	11.3–17.9
Pseudotsuga	5.7–13.7	7.1–13.9
Angiosperms	4.1–25.2	9.1–29.8
Betula	7.4–13.1	
Fagus	9.8–17.0	
Populus	8.7–22.4	

Table 10.   ANP, NPP, and growth rates for long-term tropical crops.

| | Growth Rate (g m^{-2} day^{-1}) | | ANP (NPP) |
	Maximum	Average	(tonne ha^{-1} yr^{-1})
Banana[a]	—	10.8	39.3
	—		(62.5)
Manioc[c] C$_3$	16[b]	4.5	15–18
			(30–40)
Oil palm[c] C$_3$	—	11.0	40.0
Sugar cane[c] C$_4$	38	17.8	63–67

[a] Sundarraj and Mitchell, unpublished data.
[b] Cock et al., 1979.
[c] Loomis and Gerakis, 1975.

prairies in the United States. Mixed cropping is a tradition in which farmers select and combine cultivars in response to local condition with the object of obtaining more consistent yields, spreading the labor requirements more evenly through time, reducing soil erosion, and exploiting the positive yield interactions between cultivars (Ruthenberg, 1971; Grigg, 1974; Stelly, 1976; Kowal and Kassam, 1978; Norman, 1979; Webster and Wilson, 1980; Mutsaers et al., 1981; ICRISAT, 1981). The data are inadequate for generalizations about the ANP under mixed cropping, and it may well turn out that these practices are so closely geared to local conditions that generalizations are not meaningful. At present the biological basis for the yields under mixed cropping can be reviewed.

There is a great deal of misunderstanding about the productivity of mixtures of crops. It has been presumed that mixed cropping is an inefficient folk tradition that cannot be practiced on a large scale. The charge of inefficiency is based on the assumption that the increment of labor required in mixed cropping is large relative to any gain in the yield of a mixed crop. However, recent reviews (Stelly, 1976; ICRISAT, 1981) provide evidence that has helped to dispel these notions and direct attention to this remarkable agricultural tradition. Unfortunately, information is limited to yields under mixed-cropping systems, and yield data alone do not help in explaining how the yields are affected by interactions between cultivars.

Virtually all scientific studies of productivity of crop mixtures, such as those reviewed by Trenblath (1974) and the large-scale coordinated tests carried out in India since 1972 (Indian Counc. Agric. Res., 1974–79) and the ICRISAT (1981), are limited to two-cultivar mixtures. These two-cultivar mixtures can be used to identify and illustrate the kinds of interactions found in crop mixtures and to develop and test analytical procedures.

The two-crop systems devised for research by agriculturalists fail to give any idea of the complexities common in the traditional systems of mixed cropping. Webster and Wilson (1980) describe African systems in which the spatial patterns of cultivars are modified with respect to microtopography and the planting system can include combinations of intercropping, relay cropping, and seasonal rotations that defy classification. In some of the dryer areas of India over half the plantings can involve at least four cultivars (Jodha, 1981). It is not reasonable to think that studies of two-cultivar systems can be extrapolated to explain the interactions in more complex systems. It is necessary to have independent studies of the more complex system commonly used in village-level agricultures.

Mixed cropping includes an array of practices. If the mixture of cultivars is planted at the same time (intercropping), it may involve an orderly pattern of planting or the broadcast planting of mixtures of seeds. Another common technique is to plant cultivars in a growing crop, that is, relay cropping. If the cultivars in an intercrop differ greatly in their growth rate, an intercrop may function much like a relay crop.

Although the direct evidence is lacking, I believe the mixed cropping I have seen in India and found described in the reviews mentioned above represents a very sophisticated management and exploitation of plant communities. This richly varied set of traditions has evolved through trial and error by farmers who are acutely aware of the costs and benefits. When this empirical folk science is studied and evaluated, it may well be found to be a very sophisticated kind of ecological engineering.

Trenblath (1974) used two terms to measure differences between monocrops and mixtures: relative yield total and overyield. Relative yield (RY) is redefined here as the total of the contributions of each crop in a mixture as a fraction of what the crops would yield in a monocrop. For two equally dense cultivars in a mixed crop, RY is determined from crop yields (Y) as follows:

$$RY = \frac{Y_{ij}}{Y_{ii}} + \frac{Y_{ji}}{Y_{jj}}$$

hence, if the two crops had yields reduced to half the monocrop yield, $RY = 1.0$, then the farmer would harvest as much from a field divided equally between the two cultivars in monocrops as from the field in a mixed crop. Overyield occurs when the combined yield of a mixed crop exceeds the yield of the component crop with the largest yield as a monocrop. A measure of considerable popularity is the land equivalent ratio (LER), which is the area of land needed to attain monocrop yields equivalent to that of a unit of land in a mixed crop (ICRISAT, 1981).

Both relay and mixed cropping were practiced in China 2000 years ago (Hsu, 1980) and continue to be an important component of Chinese agriculture (Buck, 1930; FAO, 1980). Soybean and maize are commonly planted as intercrops and the members of Buck's (1930) team have recorded monocrop and intercrop yields with labor inputs for several villages (Table 11). The production of the intercrop system was always higher, although climatic differences could have a profound effect on yield. From 1922 to 1923 the yield of maize in mixed crops at Yenshan fell by 20%, while the yield of intercropped soybeans rose 38%. Intercrops were superior and, additionally, the intercrop yields varied less than monocrops for these two years.

Relative labor (RL) was taken as the days ha^{-1} for the mixed crop divided by the sum of half the days ha^{-1} for each monocrop. Human labor inputs are greater for intercrops largely because of work at harvest. Labor inputs vary greatly from village to village, but it is quite clear that the increment of labor for mixed crops is small relative to the yield increase in two of the villages. There was always a substantial saving in the labor input of draft animal time relative to yield, which is extremely important because animal labor time is the common limiting resource at the preferred time of planting.

Norman (1978) factored out timing in his economic analysis of Zaria farming. The crop value relative to the labor input was 25% greater for

Table 11.   ANP, relative labor (RL), and relative yield (RY) of maize–soybean intercrops in China. Economic yield from Buck (1930) converted to ANP with h.i. for maize from Zelitch (1975) and soybean from FAO (1959).

	Monocrop Maize	Intercrop Maize – Soybean		Monocrop Soybean	RY (RL)
Yenshan 1922					
ANP (tonne ha^{-1})	2.68	1.44	1.09	1.52	1.20
Labor (hr ha^{-1})					
human	681	689		409	(1.26)
bullock	161	124		74	(1.06)
Yenshan 1923					
ANP (tonne ha^{-1})	2.17	1.08	1.46	2.11	1.16[a]
Kaifeng 1923					
ANP (tonne ha^{-1})	3.30	1.00	3.45	4.35	1.16[a]
Labor (hr ha^{-1})					
human	293	190		241	(0.71)
bullock	138	103		172	(0.66)
Kiangning (T.) 1924					
ANP (tonne ha^{-1})	5.06	3.55	3.37	4.92	1.39[a]
Labor (hr ha^{-1})					
human	640	1099		792	(1.53)
bullock	49	66		49	(1.35)

[a] Overyield.

monocrops, but the crop value relative to the labor input during June and July, when labor demands are highest, was 45% greater for intercrops. In addition to that benefit, Norman (1978) estimated that the probability of mixed crop yields exceeding that of sole crops was 0.715.

The consistency of yields is commonly used to account for the use of mixed crops. This reliability of yield was clearly demonstrated in the experiments of Rao and Willey (1978). Three combinations of cultivars were tested for yields with optimal moisture and under water stress (Table 12). On the average, water stress reduced the yield of sole crops by about 0.80 tonnes ha⁻¹, but total yield from the mixed crops fell only 0.34 tonnes ha⁻¹. In favorable seasons the yields from monocrops and intercrops were similar, but during dry years, which are the most critical times for a village, even these simple intercrops had yields well above monocrops. Hence, the value of mixed cropping is related to the way it responds to weather rather than the simple averages for relative yield and overyield. Intercropping is a kind of ecological insurance allowing farmers to achieve higher average yields in all years; this benefit is also the justification for the concentration of research on mixed cropping at ICRISAT (1981). Reliability of yield is particularly critical in nearly self-sustaining village ecosystems and less important for the interlinked farms of large-scale industrial economies.

The interactions between the cultivars in a mixture often result in a sharp reduction of the yield of one cultivar (e.g., for sorghum with groundnut, Table 12). Other kinds of interactions have been examined by Francis et al.

Table 12. Effect of water stress on the ANP and relative yield (*RY*) of intercrops. Economic yields (Rao and Willey, 1978) were converted to ANP with h.i. values of 0.26 for millet (Mitchell, 1979), 0.40 for sorghum (Zelitch, 1975), and 0.44 for peanut (FAO, 1959).

| | Yields in tonne ha⁻¹ | | | | | |
| | No Stress | | | Water Stress | | |
	Monocrop	**Intercrop**	*RY*	**Monocrop**	**Intercrop**	*RY*
Millet[a]	9.70	6.93	1.14	7.76	5.42	1.26
Groundnut	5.85	1.95		3.89	1.93	
Sorghum	9.34	4.05	0.93	6.74	3.45	1.19
Groundnut	5.85	3.05		3.89	2.90	
Millet[a]	9.70	7.72	1.09[b]	7.76	6.13	1.26[b]
Sorghum	9.33	2.72		6.74	2.98	

[a] *Elusine.*
[b] Overyield.

Table 13. ANP of a bean and maize intercrop at densities of 88,000 plants ha^{-1} in a 1:1 intercrop and 44,000 plants ha^{-1} in monocrops. Based on experiments of Francis et al. (1976). Crop growth rates were calculated for a growing season of 105 days.

| | Production in kg ha^{-1} | | | |
| | Monocrop | Intercrop | | Monocrop |
	Maize	Maize	Beans	Beans
Seeds	5.56	6.22	0.37	1.84
Pods	—	—	0.29	0.58
Stems and leaves	15.01	15.84	0.65	1.92
ANP	20.57	22.06	1.31	4.34
Crop growth rate $^{\bullet}$ (g m^{-2} day^{-1})	19.6	21.0	1.24	4.1

(1976) for maize and climbing bean (Table 13). Legume–maize mixtures are widely used in South America and elsewhere. Economic yields are increased ($RY = 1.32$) for this combination, but there are even greater differences in ANP and its components. The ANP of mixtures rises to 1.88 times the equivalent of monocrops with striking changes in the partitioning of growth. Maize h.i. remains at about 0.28 while the h.i. of bean falls from 0.42 in a monocrop to 0.28 in an intercrop.

The interactions in mixtures of cultivars are varied, and a great deal of preliminary data is given in the Proceedings of the International Workshop on Intercropping (ICRISAT, 1981). Trenblath (1974) lists 12 ways in which yield can be affected in mixtures of two cultivars. Several different responses may be found in even simple systems, and it is very likely they are all significantly involved in a complex mixed-crop system such as described in Table 14. The field was planted on June 26 near the village of Pathanamputti (about 50 miles east of Trichinapalli, Tamil Nadu, India). Regrettably, data for this system, which is typical of extensive areas of dryland agriculture in India, are limited to descriptive data for planting supplemented by basic information on the biology of the cultivars. Nearly all the dryland agriculture planted in the area was in intercrops of 5–10 cultivars, and each farmer had a clear planting scheme in mind. One set of seeds (5 cultivars, see Table 14 for relative abundances) was broadcast over the freshly plowed field. Planting furrows were cut with bullock plows. After planting five rows of red gram with a scattering of *Lablab purpureus* among them, a sixth row of caster bean was planted. The three species of *Vigna* sprouted quickly, with *V. trilobata* quickly forming a thin ground cover. The scattered cowpea and mung bean plants then added a broken canopy at about 50 cm above the ground. This was the first of what might be distinguished as a succession of three cano-

Table 14. Traits of cultivars used in an intercropping system at Puthanamputti (Tamil Nadu, India). Planting is by both broadcast and drilling with bullock-drawn plows in late June or July. The terms *food* and *fodder* are used to distinguish human consumption from cattle consumption. Original observations made with Dr. T. S. P. Sundarraj.

	Rain Required (cm)	Uses (days of harvest)		
		Seeds	Leaves	Stem
Broadcast Planting (% of seeds)				
Vigna trilobata 48% Shallow rooted, trailing stems, 25–50 cm long.[a]	25–50	Food (65–120)	Fodder[b] (65–120)	Fuel (200+)
V. unguiculata (cowpea) 6% Shallow rooted, erect to 80 cm tall.	122[a] (70–170)	Food (90–120)	Fodder[b] (90–120)	Fuel (200+)
V. radiata (mung bean) 7% Branching, 30–120 cm tall.[c]	125[a] (30–410)	Food (65–120)	Fodder[b] (65–120)	Fuel (200+)
Hybiscus cannabis 2% Tall stems up to 5 m.	50–75[a]	Oil[d] (30–60)	Food (150–180)	Fuel Fiber
Sorghum vulgare 51% Roots spreading to 1 m and down to 75 cm. 1–3 m tall.	(150–180) ?	Food (180)	Fodder (180)	Fuel (180)
Drilled in Rows				
Cajanus cajan (red gram) (5 rows); root to 2 m, bush up to 4 m.	145[a] (50–400)	Food (190–210)	Fodder (190–210)	Fuel (190–210)

Ricinus communis (caster bean) (every 6th row); roots spread to 25 cm, down to 2 m, tall.	50[b]	Oil (190–210)	—	Fuel (190–210)
Lablab purpureus (field bean) Very deep root, bush 2–3 m.	145 (70–400)	Food (120–150)	Fodder (120–150)	Fuel (200+)

[a] Duke, 1981.
[b] Sundarraj and Thulasides, 1976.
[c] Oil is not extracted presumably due to low yields.

39

pies. As the growing season progressed, the *Vigna* canopy was penetrated by the other five cultivars, and these began to shade the low *Vigna* canopy just a bit before the harvest of the three *Vigna* species. This harvest ranged in time from 60 to 120 days after planting depending on rainfall. During this time, leaves were picked from *Hybiscus* and used as a vegetable.

By the time the early beans were harvested and the canopy of the fast growing shallow rooted *Vigna* died back, a canopy of sorghum appeared. In extremely dry years or when the summer monsoon (July, August) fails, there may be no *Vigna* crop and a very stunted growth of sorghum. The local variety of sorghum tillers readily and can remain stunted but ready for a burst of growth if there is any rain. Hot, dry winds increase the effect of droughts, but the caster beans in every sixth row create tall leafy windbreaks that are thought to reduce the dessicating effects of the winds on the other crops.

The scattered lablab plants, which were sown with the red gram, are harvested shortly before the sorghum. By then red gram, which grows as a single apically dominant spike for three months, loses its apical dominance and bushes out to form still another canopy. It will bear a crop as much as 7 or 8 months after planting. The slow growing red gram plants will have a burst of growth if there is very much rain from the winter monsoons (October, November).

The farmers of Pathanamputti, like all Indian dryland farmers, must contend with extremely variable monsoons. The village receives less than the average rain for the state of Tamil Nadu. Detailed local weather data are not available, but averages for the state of Tamil Nadu indicate that half the time there will be from 70 to 95 cm rain per year (95% confidence limits are 46–120 cm). The moisture requirements of the cultivars span this range (Table 14), and the range of tolerances may well be even wider in mixtures with a sequence of canopies, nonoverlapping arrays of roots, and some protection from dessication by wind.

There is more information available for long-cycle swidden (slash and burn) intercropping, which can involve from 10 to over 50 cultivars. Rappaport (1971) described the Tsembaga gardens, which are much more elaborately managed cultivar communities than those of dryland agriculture. The transition from lower diversity mixed cropping to the high diversity of swidden agriculture can be followed in comparative studies of short- to long-term cycles in northeast India by Toky and Ramakrishnan (1981).

It seems likely that both human preferences and ecology govern the selection and combination of cultivars (Table 15). Grains may be preferred over tubers because they are less bulky and easier to store with less risk of loss. Yields are very low on these hill lands when the fallow period is only 5 years, and all measures of production increase with longer fallow periods. The difference is smallest in NPP between the 10- and 30-year cycles.

Table 15. Economic yield and community traits for swidden (jhum) agriculture in northeast India (Megalya) at cycles of 5, 10, and 30 years.[a]

	Fallow Period		
	5 yr	10 yr	30 yr
Economic yields (tonne ha^{-1} yr^{-1})			
Seeds	0.107	1.153	2.180
Leaf–fruit	0.129	0.074	0.024
Tubers	0.320	0.613	0.192
Total	0.556	1.842	3.296
ANP	14.060	11.576	15.213
Growth rate (g m^{-2} day^{-1})	3.8	3.2	4.2
NPP	18.461	14.709	17.746
Growth rate (g m^{-2} day^{-1})	5.1	4.0	4.9
Number of cultivars	8	12	14
LAI	0.59	1.49	3.20
h.i.[b]	0.030	0.125	0.135
h.i. (grain + seed)[c]	0.184	0.230	0.182
Labor (days ha^{-1} yr^{-1})	149	305	436

[a] Toky and Ramakrishnan, 1981.
[b] Crop yield/NPP because seeds, leaves, and tubers are harvested.
[c] Aboveground crop yield/ANP.

The traditional cycle has been 25–30 years, and the total gain in yield (30%) requires 43% more labor over that needed on a 10-year cycle. Total yield may be less important than the kinds of crops that can be grown. There is an 89% increase in the yield of seed crops under the long planting cycle. Erosion of fields and the loss of nutrients are related to leaf cover. There is more than a doubling of leaf cover from 10- to 30-year cycles, which will reduce the direct effect of rain in producing erosion. As indicated by changes in soil nutrients during the crop cycle (Ramakrishnan and Toky, 1981), there is less loss of nutrients from farming after a 30-year cycle. Much the same sort of yield response may account for the patterns of cultural adaptation Gross et al. (1979) described for native tribes in Brazil.

While swidden agriculture is the best known example of diverse cropping at high nutrient levels, there are a few very special high-input mixed-cropping systems that indicate the potential for mixed cropping. One of these is the betel leaf garden I have studied in collaboration with Dr. T. S. P. Sundarraj. Betel leaves (*Piper betle*) are a very profitable cash crop. They are used to wrap betel nut (*Areca catechu*) and spices, which are chewed as an astringent and digestif after meals. The vines require shade and are grown under a complex canopy of four major cultivars. The system requires heavy subsidies of water and manure. While an extreme example of complex structure, the average growth rate of 14 g m^{-2} day^{-1} is well above any crop except napiergrass. The betel groves show the extent to which ecological engineering has been developed by traditional farmers (Table 16).

The high yields are achieved because of both the inputs of water and manure, as well as the management of a three-layer canopy. At 4.5 m a constantly trimmed canopy of *Sesbania grandifolia* (1800 plants ha^{-1}) provides the top layer lying over a denser canopy at 2–3 m, which consists of three species, *S. grandifolia* (7400 plants ha^{-1}), *Erythrina indica* (3000 plants ha^{-1}), and *Moringa pterygosperma* (11,100 plants ha^{-1}). This canopy is also regulated through almost daily trimming. The trimmings from the canopy are used as human food and cattle fodder. From ground level to about 2 m, betel vines (37,000 plants ha^{-1}) are trained to grow up the stems of the canopy trees. Some 3000–6000 plants of banana, chilie, and eggplant are scattered irregularly through each hectare of a betel grove. A regular supply of very high quality cattle forage and human food is produced over much of the three-year cycle, and 25 tonnes ha^{-1} of fuel are obtained at harvest at the end of the cycle.

Betel groves require a great deal of labor, about 3400 days ha^{-1} yr^{-1}, which is about 7 times that required for year round rice culture, and the yields are 3.7 times that of rice (including the straw which is of low value). That is not a disproportionately high investment of human labor time be-

Table 16.   Average annual production from a betel garden. Except for a loss of leaves at the end of the three-year cycle, the crop yields are equivalent to ANP.

Component	Use	Growth Rate (g m^{-2} day^{-1})	Dry Weight (tonne ha^{-1})	10^6 kcal yr^{-1}
Betel vine		2.22		
Leaves	Human food		5.77	23.67
Stems	Fiber		2.32	9.63
Sesbania		8.27		
Leaves	Cattle food		9.57	42.23
Stems	Fuel		21.61	101.74
Erythrina		0.83		
Leaves	Cattle food		2.41	9.08
Stems	Fuel		.62	2.92
Moringa		2.59		
Leaves	Human, fodder		4.44	20.17
Fruits	Human food		2.22	8.53
Stems	Fuel	2.78	2.78	13.09
Total		14.18	51.74	231.06

cause the time demands are spread fairly uniformly and produce a consistent crop of high quality fodder and green vegetables throughout the year. The ANP of this managed Indian betel garden is as high as that of the most productive naturally nutrient subsidized tropical marshes (Table 4), and all the ANP is economically valuable.

DISCUSSION

Transeau's (1926) study of a cornfield is often cited as a landmark in the development of a general scheme for the analysis of primary production, but it was 20 years before ecologists began to use the ecosystem concepts implicit in Transeau's work. It was another 25 or 30 years before there was a significant extension of ecological concepts to agricultural systems. The initiative for agroecosystem research was an independent development in the emerging agronomic discipline of crop physiology, and the field has been fueled by the practical successes with high-yielding varieties. Current studies of crops as plant communities can be traced back to the theoretical models for light absorption by plant canopies (Monsi and Saeki, 1953), which paved the way for the a priori design of plant ideotypes that form efficient communities (Donald, 1968). Reviews of the way crops function at the community level in the volumes edited by Evans (1975), Monteith (1975–1976), and Alvim and Kozlowski (1977) show how a comprehensive view of the biological basis of primary production has been used to more fully exploit the potential for increasing economic yields (Donald and Hamblin, 1976; Gifford and Evans, 1981). This research on crop physiology provides the basis for the synthesis of agronomic and ecological research on primary production that was anticipated by Transeau (1926).

Primary production is the energy fixed through the process of photosynthesis. The concept of primary production as the sum of the energy released in metabolism and embodied in growth can be expressed as a simple balanced equation for the energy budget of a plant community (Table 1). Yet it is impractical, if not impossible, to obtain direct and comparable measures of the three components of the energy budget. The rates of the physiological processes, photosynthesis and metabolism, which are measured for isolated parts of plants over short periods of time (minutes) under constant conditions, cannot be extrapolated to get the cumulative value for these processes over the annual cycle of a plant community. Estimates of the growth of plant communities must be based on harvest data, and sampling periods of less than a few days are inappropriate. There is no direct way to combine the harvest data for cumulative growth over periods of 10^4 min. with primary data on physiological processes in minutes under constant conditions to obtain a complete energy budget for primary production over the annual cycle of a plant community.

The practical difficulties of estimating primary production can be ignored if it is argued that net primary production (NPP = growth) sets limits to the structure and activity of an ecosystem because NPP is the biomass that forms the structure of the plant community and supplies nutrients to herbivores and decomposers. There is a practical difficulty in estimating NPP. Aboveground net primary production (ANP) is relatively easy to estimate from harvest data, but the belowground production is extremely difficult to harvest, and root harvests are inevitably incomplete. Nearly all estimates of NPP use a root : shoot ratio to convert ANP to NPP. All the available data suggests that there is no consistent relationship between root and shoot production.

ANP is the only body of reasonably accurate primary data for production that is extensive enough for comparative studies, and these data are still limited to spot records with no information on year-to-year variability in natural ecosystems. There are few studies of ANP for agroecosystems, but each cultivar has an innate yield : shoot ratio that can be used to obtain reliable estimates of ANP from data on the economic yields of crops.

Despite all the practical problems that limit data on production, the data on the ANP of natural systems do follow patterns that correlate closely with temperature and measures of water availability (Fig. 1). This correlation of ANP with climate justifies the view that ANP reflects biological responses general enough to be the basis for comparative studies of natural and agricultural primary production. Several questions can be addressed through such comparisons: (1) What is the potential productivity of a plant community? (2) Does the ANP of natural systems differ from that of comparable crop systems? (3) If there are differences, are they the result of differences in the environment (disturbed ground and chemical inputs, for example) or are biological factors responsible for the differences?

Large-scale differences in ANP appear to be responses to climate. Obviously, the major differences in Figure 1 are caused by differences in the length of the growing season. Seasonal effects can be factored out by using the growth rate in g m^{-2} day^{-1} to see if the growth rates are independent of the length of the growing season. Average growth rates are probably systematically affected by growth patterns, such as perennials versus annuals. When growth has been recorded for reasonably short periods of time (usually a week), the maximum growth rate can be distinguished. This is the best basis for estimating the biological capacity for production by a plant community independent of both the length of the growing season and limits imposed by the patterns of growth. Growth rates based on shoot biomass are a fraction of the total growth, and that fraction must vary with ontogenetic changes in the partitioning and production of photosynthates. Most of the root growth of annuals occurs early in the growing season and is thought to usually decline to quite low levels when flowering and seed set occurs. If this is true, then growth rates based on shoot growth may happen to approach the total growth (NPP) at the time of seed set. Peak values for shoot growth rates reach 50–55 g m^{-2} day^{-1} for a few days or weeks during or before seed set in a wide variety of crops and natural communities (Tables 4, 6, and 7). Because so many of the maximum values fall in this narrow range, and there is good reason to believe these measures reflect total growth, it is reasonable to ask if they represent a biological limit for the growth rate of plant communities.

These maximum growth rates can be compared to the theoretical limits for photosynthesis. If the estimates are standardized for photosynthesis with an input of 15 MJ m^{-2} day^{-1} (the average daily input from May through August in the American midwest) and a coefficient of 17 kj g^{-1} for biomass, the estimate of Loomis and Williams (1963) will give 71 g m^{-2} day^{-1}. Monteith's (1977) formulas give estimates of 91 g m^{-2} day^{-1} for C$_3$ plants and 125 for C$_4$ plants.

The canopies of plant communities cannot be as efficient in absorbing light as the ideal surfaces assumed in the models, and the daily flux of light, temperature, and humidity will rarely provide the ideal conditions necessary to reach the maximum photosynthetic rate. Given these barriers to achieving the theoretical limit in the real world, values of 50–55 g m^{-2} day^{-1} in nature, from 44 to 77% of the physiological limit for photosynthesis under optimal temperatures with no water limitation may be close to the maximum possible.

The photosynthetic system might systematically affect realized growth rates. C$_4$ plants are adapted to deal with water stress at high temperatures, which occur in climates with higher levels of solar radiation, while C$_3$ plants are more efficient at lower temperatures. Keulin et al. (1976) argue: "The very large potential advantage of the C$_4$ mechanism . . . is progressively attenuated in moving from the microscopic to the macroscopic parameters until, at the level of crop growth rate, there is no apparent difference between the best of the two examples when grown in their own preferred environments." C$_4$ plants do reach the highest levels of production, but these are exceptions. The very broad overlap of C$_3$ with C$_4$ plants (Tables 4, 6, and 7) implies that there is no general and consistent difference associated with the photosynthetic system at the level of community production.

Growth rates of 50 g m^{-2} day^{-1} would give a production of 182 tonnes ha^{-1} yr^{-1} for the continuous growth of a plant community maintaining a maximally efficient canopy under a favorable year around climate. Plant communities change through time and so does climate. The record for ANP is achieved in intensely managed stands of napiergrass, *Pennesetum purpu-*

reum, that produce 85 tonnes ha^{-1} yr^{-1} (Table 6). The cycles of the most efficient crops range from 100 to 150 days with ANP peaking at 20–30 tonnes ha^{-1} (Tables 4 and 7) so their rates of growth are equivalent to those of *Pennesetum*. Average rates of growth based on ANP over the entire crop cycle with the attendant stresses of temperature and moisture are close to being 25% of the theoretical maximum based on total production. Average growth rates for total growth may well approach the theoretical maximum.

There are no exact matches of agricultural and natural communities on which to base comparisons. The higher values for crop yields are from areas where water is not seriously limiting and nutrients are subsidized, so it is appropriate to compare crops with marshes and floodplains that receive natural nutrient and water subsidies. Such natural communities attain higher levels of productivity than reached with organic agriculture. With high chemical nutrient subsidies, crop production may equal that of naturally subsidized marshes. The maxima for the ANP of temperate marshes, forests, and chemically subsidized agriculture generally fall between 20 and 30 tonnes ha^{-1} yr^{-1} (Tables 4, 7, and 9). The claims of agricultural production exceeding natural production usually appear to involve inappropriate comparisons of crops that are given subsidies of nutrients or water or both with adjacent grassland communities that are not nutrient subsidized.

The correlation of ANP with temperature and rainfall (Fig. 1) is the basis for a number of models for predicting agricultural production from temperature and rainfall (e.g., Thompson, 1975). Given the possibility that communities may be close to the theoretical maximum, these patterns might reflect predictable biological responses to climate. Models for predicting ANP have been most completely explored by crop physiologists. These range from functional models based on the physiological determinants of the efficiency of light fixation to the processes that determine the partitioning of photosynthates over ecological time. These are the most realistic models, but while they are useful in examining short-term physiological processes, they are extremely difficult to use over ecological time. The difficulty encountered in the use of biologically realistic models may be due to the fundamental problems encountered in extrapolating from physiological time, rates per minute under constant conditions, to the outcome over the growing season (10^5 min). Models accounting for the flux of energy through a canopy and the complex feedback loops that control and regulate physiological processes throughout the life cycle of a plant are not practical at this time (Sheehy and Cooper, 1973; Monteith, 1977; Wright and Keener, 1982). The models that are the most practical and reliable over ecological time periods are based on correlations of production with temperature and some measure of water resources (Fig. 2). The agronomic models based on these correlations are similar to the models ecologists use to account for the patterns of primary production on a global scale.

The increases in economic yield are not accompanied by an increase in ANP but rather reflect a change in the partitioning of ANP with the portion going into grain increasing from levels of 25–30% to as much as 55% in the major food grains. The most spectacular increases are the result of the *a priori* design of non-competitive phenotypes that form stable canopies with an optimal leaf array. Some of the increases in ANP may be attributed to an extension of the period of maximum crop growth rate. The experiments of agronomists have done much to clarify the role of intraspecific competition in plant communities. The high-yielding varieties are ideotypes designed to be non-competitive so that more of the photosynthate can be transferred to grain yields (Figs. 3 and 4, Table 8).

In accounting for the biological interactions that affect production, ecologists and agronomists must learn more about the year-to-year responses of plant communities to climate. It is commonly argued that the yields of monocrops must vary more with climate than complex assemblages of species in natural communities. Mixed-cropping systems seem to offer an unusually clear basis for testing this intuitive presumption of a relation between variability of production and community structure.

The traditional mixtures of several cultivars are claimed to provide a less variable yield than monocrops of the cultivars used in mixtures, and mixtures do appear to have more stable yields (Table 12). Comparisons of monocrops with the behavior of cultivars in mixtures may be the best way to come to grips with the nature of the interactions in complex plant communities (Tables 12–16). Agronomists are now convinced that mixed cropping does give larger and more dependable yields than monocropping, at least in dryland farming (ICRISAT, 1981). The advantages of mixed cropping can be gained only with hand labor because each of the cultivars of the mixture must be planted and harvested individually. Scientific experimentation has been limited to two-way interactions, which are extremely complex, but these do not give much insight into the even more complex and diverse mixtures that are extremely common in traditional mixed cropping. A study of these traditions would seem likely to reveal fundamental information on the way interactions in complex canopies affect yields and variability of yields, as well as testing the beliefs of farmers who practice this complex folk technology.

ACKNOWLEDGMENTS

My interest in agroecosystem research is the outcome of working to develop appropriate materials for teaching ecology in India while on a Fulbright Professorship to India (1976–1977) and the continuing encouragement of many friends in India. Here at Ohio State University I have received much advice from Carroll Swanson and helpful criticism from Ralph Boerner, Gary Middelbach, and Kay Gross.

REFERENCES

Alvim, P. de T. and Kozlowski, T. T. (1977). *Ecophysiology of Tropical Crops.* Academic Press, New York.

Austin, R. B. (1982). Crop characteristics and the potential yield of wheat. *J. Agric. Sci. Camb.* **98:**447–453.

Austin, R. B., Bingham, J., Blackwell, R. D., Evans, L. T., Ford, M. A., Morgan, C. L., and Taylor, M. (1980). Genetic improvements in winter wheat yields since 1900 and associated physiological changes. *J. Agric. Sci. Camb.* **94:**675–689.

Bath, B. H. S. van (1963). *The Agrarian History of Western Europe.* A.D. *500–1800.* E. Arnold Publishers, London.

Birch, J. B. and Cooley, J. L. (1982). Production and standing crop patterns of giant cutgrass (*Zizaniopsis miliacea*) in a freshwater tidal marsh. *Oecologia* **52:**230–235.

Brinson, M. M., Lugo, A. E., and Brown, S. (1981). Primary productivity, decomposition and consumer activity in freshwater wetlands. *Ann. Rev. Ecol. Syst.* **12:**123–161.

Buck, J. L. (1930). *Chinese Farm Economy.* University of Chicago Press, Chicago.

Cannell, M. G. R. (1982). *World Forest Biomass and Primary Production Data.* Academic Press, London.

Chandler, R. F. (1979). *Rice in the Tropics.* Westview Press, Boulder, Colorado.

Cock, J. H., Franklin, D., Sandoval, G., and Jori, D. (1979). The ideal cassava plant for maximum yield. *Crop Sci.* **19:**271–279.

Cooper, J. P. (1975). Control of photosynthetic production in terrestrial systems. In *Photosynthesis and Productivity in Different Environments.* J. P. Cooper (ed.), Cambridge University Press, Cambridge, pp. 593–621.

Coupland, R. T. (1979). Conclusion. In *Grassland Ecosystems of the World.* J. P. Cooper (ed.), Cambridge, London, pp. 335–355.

Donald, C. M. (1968). The breeding of plant ideotypes. *Euphytica* **17:**385–403.

Donald, C. M. and Hamblin, J. (1976). The biological yield and harvest index of cereals as agronomic and plant breeding criteria. *Adv. Agron.* **28:**361–405.

Duke, J. L. (1981). *Handbook of Legumes of World Economic Importance.* Plenum, New York.

Duncan, W. G. (1975). Maize. In *Crop Physiology.* L. T. Evans (ed.), Cambridge University Press, Cambridge, pp. 23–50.

Dunstone, R. L., Gifford, R. M., and Evans, L. T. (1973). Photosynthetic characteristics of modern and primitive wheat species in relation to ontogeny and adaptation to light. *Austr. J. Biol. Sci.* **26:**295–307.

Eastin, J., Haskins, F. A., Sullivan, C. Y., and van Bavel, C. H. M., (1969). *Physiological Aspects of Crop Yield.* American Society of Agronomy, Madison, Wisconsin.

Elston, J. F., Greenland, D. J., and Dennett, M. D. (1980). Long term trends in the aerial and edaphic environment. In *Opportunities for Increasing Crop Yields*. R. G. Hurd, P. V. Briscoe, and C. Dennis (eds.), Pitman, Boston, pp. 87–96.

Evans, L. T. (1975). *Crop Physiology*. Cambridge University Press, Cambridge.

Evans, L. T. and Dunstone, R. L. (1970). Some physiological aspects of evolution in wheat. *Austr. J. Biol. Sci.* **23**:725–41.

Falk, J. H. (1980). Primary productivity of lawns in temperate environment. *J. Appl. Ecol.* **17**:689–696.

FAO (1959). Tabulated information on tropical and subtropical grain legumes. Food Agriculture Organization, Rome.

FAO (1980). China: Multiple Cropping and Related Crop Production Technology. Food and Agricultural Organization. Plant Product. and Protect. Paper No. 22, 57 pp.

Francis, C. A., Flor, C. A., and Temple, S. R. (1976). Adapting varieties for intercropping systems in the tropics. In *Multiple Cropping*. M. Stelly (ed.), Spec. Publ. 27. American Society of Agronomy, Madison, Wisconsin, pp. 235–253.

Gaudet, J. J. (1977). Uptake, accumulation and loss of nutrients by papyrus in tropical swamps. *Ecol.* **58**:415–422.

Gifford, R. M. (1974). A comparison of potential photosynthesis, productivity and yield of plant species with different photosynthetic metabolism. *Aust. J. Plant Physiol.* **1**:107–117.

Gifford, R. M. and Evans, L. T. (1981). Photosynthesis, carbon partitioning and yield. *Ann. Rev. Plant Physiol.* **32**:485–509.

Giurgevich, J. R. and Dunn, E. L. (1982). Seasonal patterns of daily net photosynthesis, transpiration, and net primary productivity of *Juncus roemerianus* and *Spartina alterniflora* in a Georgia salt marsh. *Oecologia* **52**:404–410.

Good, R. E. and Good, N. F. (1975). Vegetation and production of the Woodbury Creek Hessian Run freshwater tidal marshes. *Bartonia* **43**:38–45.

Griffing, B. (1967). Selection in reference to biological groups. I. Individual and group selection applied to populations of unordered groups. *Austr. J. Biol. Sci.* **20**:127–139.

Grigg, D. B. (1974). *The Agricultural Systems of the World*. Cambridge University Press, Cambridge.

Gross, D. R., Eilen, G., Flowers, N. M., Leoi, F. M., Ritler, M. L., and Werner, D. W. (1979). Ecology and acculturation among native peoples of central Brazil. *Science* **206**:1043–1050.

Hsu, Cho-yun. (1980). *Han Agriculture*. University of Washington Press, Seattle, Washington.

ICRISAT (1981). *Proceedings of the International Workshop on Intercropping*. Int. Crop. Res. Inst. Semi-Arid Tropics. Patancheru, India.

Indian Counc. Agric. Res. (1974–1979). All India agronomic research project. Annual Reports. 1972-3: 568 pp. (1974); 1973-4: 199 pp. (1975); 1974-5: 349 pp. (1977); 1975-6: 389 pp. (1978); 1976-7: 150+316+189 pp. (1978); 1977-8: 556 pp. (1979). New Delhi, India.

Ishizuka, Y. (1969). Engineering for higher yields. In *Physiological Aspects of Crop Yield*. J. Eastin, F. A. Haskins, C. Y. Sullivan, and C. H. M. van Bavel (eds.), American Society of Agronomy, Madison, Wisconsin, pp. 15–25.

Ishizuka, Y. (1971). Physiology of the rice plant. *Adv. Agron.* **23**:241–315.

Jennings, P. R. and de Jesus, J., Jr. (1968). Studies on competition in rice. I. Competition in mixtures of varieties. *Evolution* **22**:119–124.

Jennings, P. R. and Herrera, R. M. (1968). Studies on competition in rice. II. Competition in segregating populations. *Evolution* **22**:332–336.

Jennings, P. R. and Aquino, R. C. (1968). Studies on competition in rice. III. The Mechanism of competition among phenotypes. *Evolution* **22**:529–542.

Jodha, N. S. (1981). Resource base as a determinant of cropping patterns. In *Cropping Systems Research and Development for the Asian Farmer*. International Rice Research Institute, Los Baños, Philippines, pp. 106–126.

Keulin, H. van, Wit, C. T. de, and Lof, H. (1976). The use of simulation models for productivity studies in arid regions. In *Water and Plant Life*. O. L. Lange, L. Kappen, and E.-D. Schulze (eds.), Springer, Berlin, pp. 408–431.

Kira, T. (1975). Primary production of forests. In *Photosynthesis and Productivity in Different Environments*. J. P. Cooper (ed.), Cambridge University Press, Cambridge, pp. 5–40.

Kowal, J. M. and Kassam, A. H. (1978). *Agricultural Ecology of the Savanna*. Oxford, London.

Lee, E. (1979). Egalitarian peasant farming and rural development: The case of South Korea. In *Agrarian Systems and Rural Development*. D. Ghai, A. R. Khan, E. Lee, and S. Radwan (eds.), Hohurs and Meir Publishers, New York.

Leith, H. (1975). Modeling the primary production of the world. In *Primary Production of the Biosphere*. H. Leith and R. H. Whittaker (eds.), Springer, New York, pp. 238–263.

Leith, H. (1976). The use of correlation models to predict primary productivity from precipitation or evapotranspiration. In *Water and Plant Life*. O. L. Lange, L. Kappen, and E.-D. Schulze (eds.), Springer, Berlin, pp. 393–407.

Lennard, R. V. (1932). English agriculture under Charles II. *Econ. Hist. Rev.* **4**:23–45.

Lindeman, R. L. (1942). The trophic dynamic aspect of ecology. *Ecology* **23**:399–418.

Lockeretz, W., Shearer, G., and Kohl, D. H. (1981). Organic farming in the corn belt. *Science* **211**:540–553.

Loomis, R. S. and Gerakis, P. A. (1975). Productivity of agricultural ecosystems. In *Photosynthesis and Productivity in Different Environments*. J. P. Cooper (ed.), Cambridge University Press, Cambridge, pp. 145–172.

Loomis, R. S. and Williams, W. A. (1963). Maximum crop productivity: an estimate. *Crop Sci.* **3**:67–72.

Loomis, R. S., Williams, W. A., and Hall, A. E. (1971). Agricultural productivity. *Ann. Rev. Plant Physiol.* **22**:431–468.

Loucks, O. L. (1977). Emergence of research on agro-ecosystems. *Ann. Rev. Ecol. Syst.* **8**:173–192.

Lugo, A. E. and Brinson, M. M. (1978). Calculations of the value of salt water wetlands. In *Wetland Functions and Values: The State of Our Understanding*. P. E. Greesor, J. R. Clark, and J. E. Clark (eds.), American Water Resources Association, Minneapolis, Minnesota, pp. 120–130.

McNaughton, S. J. (1966). Ecotype function in the *Typha* community-type. *Ecol. Mon.* **36**:297–325.

Miracle, M. P. (1967). *Agriculture in the Congo Basin.* University of Wisconsin Press, Madison, Wisconsin.

Mitchell, R. (1979). *The Analysis of Indian Agro-Ecosystems.* Interprint, New Delhi.

Moerman, M. (1968). *Agricultural Change and Peasant Choice in a Thai Village.* University of California Press, Berkeley.

Monsi, M. and Saeki, T. (1953). Uber den Lichtfaktor in den Pflanzengesellschaften und seine Bedeuting for die Stoffproduktion. *Jap. J. Bot.* **14:**22–52.

Monteith, J. L. (1969). Light interception and radiative exchange in crop stands. In *Physiological Aspects of Crop Yield.* J. Eastin, F. A. Haskins, C. Y. Sullivan, and C. H. M. van Bavel (eds.), American Society of Agronomy, Madison, Wisconsin, pp. 89–113.

Monteith, J. L. 1975–1976. *Vegetation and the Atmosphere.* Vol. I (1975); Vol. 2 (1976). Academic Press, New York.

Monteith, J. L. (1977). Climate. In *Ecophysiology of Tropical Crops.* P. de T. Alvim and T. T. Kozlowski (eds.), Academic Press, New York, pp. 1–27.

Murata, Y. and Matsushima, S. (1975). Rice. In *Crop Physiology.* L. T. Evans (ed.), Cambridge University Press, Cambridge, pp. 73–99.

Murphy, P. G. (1975). Net primary productivity in tropical terrestrial ecosystems. In *Primary Productivity of the Biosphere.* H. Leith and R. H. Whittaker (eds.), Springer, New York, pp. 217–231.

Mutsaers, H. J. W., Mbouemboue, P., and Boyomo, M. (1981). Traditional food crop growing in the Yaounde area (Cameroon). Part II. Crop associations, yields, and fertility aspects. *Agro-Ecosystems* **6:**289–303.

Norman, D. N. (1978). Farming systems and the problems of improving them. In *Agricultural Ecology of the Savanna.* J. M. Kowal and A. H. Kassam (eds.), Oxford, London, pp. 318–347.

Norman, M. J. T. (1979). *Annual Cropping Systems in the Tropics.* University of Florida Press, Gainesville, Florida.

Odum, E. P. (1975). *Ecology: The Link Between the Natural and Social Sciences.* 2nd ed., Holt Rinehart and Winston, New York.

Ovington, J. D., Heitkamp, D., and Lawrence, D. B. (1963). Plant biomass and productivity of prairie, savannah, oakwood, and maize field ecosystems in central Minnesota. *Ecology* **44:**52–63.

Pomeroy, L. R., Darley, W. M., Dunn, E. L., Gallagher, J. L., Haines, E. B., and Whitney, D. M. (1981). Primary production. In *The Ecology of a Salt Marsh.* L. R. Pomeroy and R. G. Wiegert (eds.), Springer, New York, pp. 39–67.

Ramakrishnan, P. S. and Toky, O. P. (1981). Soil nutrient status of hill agroecosystems and recovery pattern after slash and burn agriculture (jhum) in northeastern India. *Plant and Soil* **60:**41–64.

Rao, M. R. and Willey, R. W. (1978). Current status of intercropping research and some suggested experimental approaches. In *Proceedings: Second Review Meeting INPUTS Project.* S. Ahmed, H. P. M. Gunasena (eds.), East–West Center, Honolulu, Hawaii, pp. 123–137.

Rappaport, R. A. (1971). The flow of energy in an agricultural society. *Sci. Amer.* **225:**116–132.

Rawski, E. S. (1972). *Agricultural Change and the Peasant Economy of South China.* Harvard University Press, Cambridge.

Richardson, C. J. (1978). Primary productivity values in fresh water wetlands. In *Wetland Functions and Values: The State of Our Understanding.* P. E. Greeson, J. R. Clark, and J. E. Clark (eds.), Amer. Water Resources Assoc., Minneapolis.

Rodin, L. E., Bazilevich, N. I., and Rozov, N. N. (1975). Productivity of the world's main ecosystems. In *Productivity of World Ecosystems.* National Academy of Science, Washington, D.C., pp. 13–26.

Ruthenberg, H. (1971). *Farming Systems in the Tropics.* Clarendon Press, Oxford.

Sheehy, J. E. and Cooper, J. P. (1973). Light interception, photosynthetic activity, and crop growth rate in canopies of six temperate forage grasses. *J. Appl. Ecol.* **10:**239–250.

Sims, P. L. and Coupland, R. T. (1979). Producers. In *Grassland Ecosystems of the World.* R. T. Coupland (ed.), Cambridge, London, pp. 49–72.

Singh, G. and Chancellor, W. (1974). Relations between farm mechanization and crop yield for a farming district in India. *Trans. Am. Soc. Agric. Eng.* **17:**808–813.

Singh, G. and Chancellor, W. (1975). Energy inputs and agricultural production under various regimes of mechanization in northern India. *Trans. Am. Soc. Agric. Eng.* **18:**252–259.

Singh, J. S., Singh, K. P., and Yadova, P. S. (1979). Ecosystem synthesis. In *Grassland Ecosystems of the World.* R. T. Coupland (ed.), Cambridge, London.

Spedding, C. R. W. (1975). *The Biology of Agricultural Systems.* Academic Press, London.

Stanhill, G. (1981). The Egyptian agro-ecosystem at the end of the eighteenth century—An analysis based on the "Description de l'Egypte." *Agro-Ecosystems* **6:**305–314.

Stelly, M. (ed.) (1976). *Multiple Cropping.* Spec. Publ. 27, American Society of Agronomy, Madison, Wisconsin.

Sundarraj, D. D. and Thulasides, G. (1976). *Botany of Field Crops.* MacMillan India, New Delhi.

Thompson, L. M. (1975). Weather variability, climatic change, and grain production. *Science* **188:**535–541.

Toky, O. P. and Ramakrishnan, P. S. (1981). Cropping and yields in agricultural systems of the northeastern hill region of India. *Agro-Ecosystems* **7:**11–25.

Transeau, E. N. (1926). The accumulation of energy by plants. *Ohio J. Sci.* **26:**1–10.

Trenblath, B. R. (1974). Biomass productivity of mixtures. *Adv. Agron.* **26:**177–210.

Turitzen, N. and Drake, B. G. (1981). Canopy structure and the photosynthetic efficiency of a C_4 grass community. In *Photosynthesis.* Vol VI. G. Akoyunoglou (ed.), Balaban Int. Sci. Services, Philadelphia, pp. 73–80.

Uchijima, I. (1976). Maize and rice. In *Vegetation and the Atmosphere.* Vol. 2. J. L. Monteith (ed.), Academic Press, New York, pp. 33–64.

USDA 1972–1981. Agricultural Statistics. United States Department of Agriculture, U.S. Government Printing Office, Washington, D.C.

Webb, W. L., Lavenroth, W. K., Szarek, S. R., and Kinerson, S. (1983). Primary production and abiotic controls in forests, grasslands, and desert ecosystems in the United States. *Ecology* **64:**134–151.

Webster, C. C. and Wilson, P. N. (1980). *Agriculture in the Tropics.* Longman, London.

Westlake, D. F. (1963). Comparisons of plant productivity. *Biol. Rev.* **38**:385–429.

Westlake, D. F. (1975). Primary production of freshwater macrophytes. In *Photosynthesis and Productivity in Different Environments.* J. P. Cooper (ed.), Cambridge University Press, Cambridge, pp. 189–206.

Whittaker, R. H. (1975). *Communities and Ecosystems.* 2nd ed. MacMillan, New York.

Whittaker, R. H. and Marks, P. L. (1975). Methods of assessing terrestrial productivity. In *Primary Production of the Biosphere.* H. Leith and R. H. Whittaker (eds.), Springer, New York, pp. 55–118.

Wiebe, G. A., Petr, F. C., and Stevens H. (1963). Interplant competition between barley genotypes. In *Statistical Genetics and Plant Breeding.* W. D. Hanson and H. F. Robinson (eds.), National Academy of Science, National Research Council Publ. No 982. Washington, D.C.

Wortman, S. (1980). World food and nutrition: The scientific and technological base. *Science* **209**:157–164.

Wright, A. D. and Keener, M. E. (1982). A test of a maize growth and development model, CORNF. *Agric. Syst.* **9**:181–197.

Yoshida, S. (1977). Rice. In *Ecophysiology of Tropical Crops.* P. de T. Alvim and T. T. Koslowski (eds.), Academic Press, New York, pp. 57–87.

Zelitch, I. (1975). Improving the efficiency of photosynthesis. *Science* **188**:626–633.

Consumers in Agroecosystems: A Landscape Perspective

John R. Krummel and Melvin I. Dyer

Environmental Sciences Division
Oak Ridge National Laboratory
Oak Ridge, Tennessee

INTRODUCTION

Agroecosystems are altered natural systems managed to enhance the productivity of a selected group of producers or consumers. Indigenous plants and animals are removed or destroyed and replaced by a few species (Pimentel, 1977). Food and fiber production both alters the natural food web structure and constrains future producer/consumer interactions in the managed agricultural system. Consumers, as both pests and products, must be managed within the constraints set by each particular production system.

One of the basic organizing forces in any ecological system is the link between production and consumption (Odum, 1971; Ricklefs, 1979; McNaughton and Wolf, 1979). Trophic-level interactions can be studied by examining food web systems that emphasize process functions (Goh, 1980; Pimm, 1982). Emphasis can be placed on processes related to physiology,

---

Research sponsored by the Ecological Research Division, Office of Health and Environmental Research, U.S. Department of Energy under contract W-7405-eng-26 with Union Carbide Corporation.

energy flow, growth and development, nutrient cycling, and birth and death rates. However, spatial dynamics of trophic-level interactions can often be ignored. Considerable theoretical work (Cohen, 1970; Horn and MacArthur, 1972; Steele, 1974; Levin, 1976; Clark et al., 1978) demonstrates the effect of spatial heterogeneity in maintaining a food web structure. The influence of spatially dependent ecological events, such as large herbivore grazing patterns, predator and prey refugia, and species and genetic diversity are central to the maintenance of ecological systems (Huffaker, 1958; Holling, 1966; May, 1975). An examination of the effect of spatial heterogeneity on producer/consumer interactions in agroecosystems can incorporate pattern and process factors that operate across conventional ecosystem boundaries (Wilson and Willis, 1975).

In this chapter we link management practices, the spatial heterogeneity component of landscape pattern, and producer–consumer interactions across an array of agricultural production systems. This linkage incorporates change in the natural landscape pattern as a function of the intensity of agroecosystem management and level of resource inputs. The degree to which the natural landscape pattern is altered will determine the relative contribution of process and pattern functions necessary to manage agricultural food webs. It is our hope that this format provides a suitable framework to address the complex array of consumer/producer interactions found in agricultural systems.

PATTERN

The basic element of the managed landscape in much of the United States is that of patches of natural vegetation scattered within a matrix of different agroecosystems (Forman and Godron, 1981; Burgess and Sharpe, 1981). Since producers and consumers form coevolutionary links under natural conditions (Ehrlich and Raven, 1969), changes in natural environmental patterns could influence the type of producer and consumer complexes that develop in agroecosystems.

Pattern in ecological systems incorporates aspects of structure and design. The physical form a vegetation complex develops over time defines the structure of the system. Grasslands, shrub communities, and temperate forests display different structural properties based on the dominant plant populations (Whittaker, 1975). Spatial heterogeneity as a component of landscape pattern can be viewed in the context of design. By design we mean the spatial mix of producer and consumer populations within defined regions. For example, remote-sensing techniques would reveal a pattern of vegetation types (i.e., deciduous forest, evergreen forest, grazing land, and wetlands) over a sampled landscape in western Kentucky (Krummel et al., 1983), whereas a similar study in the high plains would reveal a mosaic of grazed shortgrass prairie broken by riparian vegetation along streams or patches of relict forest cover along north-facing slopes.

In the absence of anthropogenic disturbances, ecological patterns develop as a function of stochastic perturbations (Shugart and West, 1977; Paine, 1966), physical constraints (Holdridge, 1967), and biological interactions (Steele, 1974). Levin (1976) identifies three biotic and abiotic factors that cause observed patterns on landscapes: (1) local uniqueness, (2) phase differences caused by uneven stages of development or recovery from perturbation, and (3) differential dispersal of organisms. Over a large area, patterns appear as a matrix of producers and consumers modified by climate and physiographic features (Whittaker, 1953; Beals, 1969).

PROCESS

While pattern variables quantify the structure and design of an ecological system, process implies functional relationships between and within the biotic and abiotic components. Indeed, classical ecosystem studies focus primarily on empirical and modeling analyses of process functions (Van Dyne, 1966). Within agricultural systems, processes include enhanced productivity of producers through fertilization, improved productivity of dairy herds through selective breeding practices, and the management of crop pests with chemical control (Spedding, 1975). Process functions incorporate parameters and variables that quantify physiology, energy flux, and nutrient cycling (Shugart and O'Neill, 1979). In the context of process functions, the management goal for animal products would be to enhance secondary productivity through increased flux of energy and nutrients. For pests that divert energy and nutrients away from the system, a management response would address the process functions (e.g., chemical control of birth and death rates of the pest complex (Krummel and Hough, 1980).

PROCESS AND PATTERN ALONG A MANAGEMENT-LANDSCAPE CONTINUUM

Because agricultural systems (defined to also include timber production) occupy about three-quarters of the land area of the United States (USDA 1982), we wish to establish a framework for consumer management based on the interrelationships between process and pattern at the landscape scale. We use a hierarchical viewpoint (Allen and Starr, 1982) to aggregate process and pattern variables in ecological systems (Fig. 1). Within this organizational hierarchy, process functions originate at a physiological level. Based on the problems being analyzed, these processes can be aggregated at the ecosystem and the biosphere level of the hierarchy. Populations form the foundations of the pattern hierarchy. The interactions of populations with the physical environment and each other combine to develop aggregations of the biota at the community, landscape, and geographic province level of the hierarchy. Every level in the process and pattern hierarchy (e.g., physiological processes, populations, ecosystems, and landscapes) reflects unique ecological properties within the context of a defined problem. For example, an analysis of the effects of insecticides on crop pests might concentrate on chemical transport across the insect cuticle (physiology in the process hierarchy), while grazing management of western rangeland would involve parameters related to species composition in the plant community (communities in the pattern hierarchy). Also, when one is interested in problems that occur over large land areas (i.e., the effect of monocultures on the origins of pest problems), one ought to develop management approaches that incorporate factors related to both landscapes and ecosystems. This is because the combination of landscape and ecosystem implies a sufficiently large spatial scale to address both pattern and process interactions. In this manner an appropriate management approach can effectively link the pro-

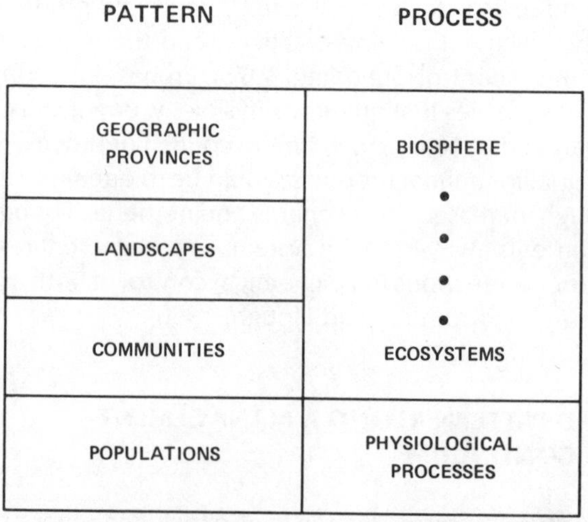

PATTERN	PROCESS
GEOGRAPHIC PROVINCES	BIOSPHERE
LANDSCAPES	• • • •
COMMUNITIES	ECOSYSTEMS
POPULATIONS	PHYSIOLOGICAL PROCESSES

Figure 1. A hierarchical view of pattern and process variables and parameters in ecological systems. Process describes functional relationships while pattern captures the spatial aspects of ecological events. Movement up the hierarchy implies increases in scale and aggregation of variables from lower levels. A line is not drawn between ecosystems and the biosphere in order to indicate a change in spatial scale rather than different functional relationships. See text for further explanation.

cesses and patterns of the ecological system to the desired management goals.

Figure 2 illustrates the link between management practices and the spatial heterogeneity component of landscape pattern. Labor, energy, machinery, fertilizer, pesticide, and other resource inputs define the level of a particular management strategy or practice (Pimentel et al., 1973; Spedding, 1975). The total level of resource inputs determines the intensity of agroecosystem management. Management intensity can be viewed as a continuum that ranges from high levels of resource inputs to little or none. Within our conceptual framework spatial heterogeneity within landscape pattern is a function of management intensity (Fig. 2). In highly managed systems with many resource inputs (i.e., corn production in the U.S. corn belt), landscape pattern becomes static and spatially homogeneous with the potential of coarse-grained utilization of resources by consumers. Less managed agroecosystems (i.e., rangeland grazing or commercial forestlands) result in relatively dynamic, spatially heterogeneous landscape patterns with an op-

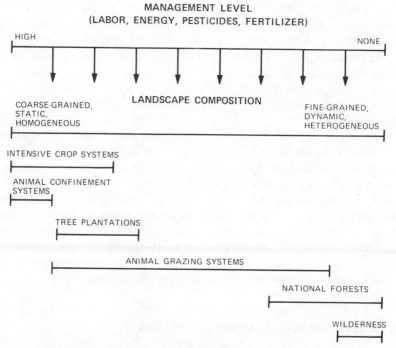

Figure 2.   Management inputs in agricultural systems range from high to none. This results in a corresponding relationship with landscape composition in each particular type of agricultural system. For example, high management inputs result in the coarse-grained, static, and homogeneous landscape of intensive crop systems. The inverse is true in wilderness systems.

tion of fine-grained utilization of a mixed resource base by consumers. For broadly defined agricultural production systems, this management–landscape continuum ranges from intensive row crops to management of native herbivores in wilderness areas (Fig. 2).

The terms *coarse-grained* and *fine-grained* apply to the orientation of the consumer in the agricultural system (MacArthur and Levins, 1964). Coarse-grained consumer populations tend to discriminate among available resources and specialize on a single uniform resource base. Whereas, fine-grained consumer populations utilize a broad resource base in the proportion at which the resources occur and thus are more general in the exploitation of a group of resources. The terms *dynamic, static, heterogeneous,* and *homogeneous* apply to the temporal and spatial mix of producers (vegetation). One would define this mix of producers through mathematical, statistical, or empirical analyses of successional events and the environmental patchiness in a landscape (Drury and Nisbet, 1973; Pielou, 1977; Forman and Godron, 1981). For example, over large geographic areas, remote-sensing data would reveal that the central Illinois portion of the Corn Belt is vegetatively more homogeneous than sheep production areas in Rocky Mountain national forestlands. At a smaller spatial scale, a monoculture of corn would be more homogeneous than a complex interplanting of crops within an old field (Root, 1973). "Dynamic" and "static" refer to the successional events of the vegetation complex. Thus, we would define continuous corn production in the Corn Belt as static, while natural vegetation undergoes dynamic changes due to stochastic events (i.e., fire) and competitive interactions (i.e., shade-tolerant versus shade-intolerant tree species).

We analyze the agricultural systems along this continuum for the relative contribution of process and pattern parameters in the management of consumers as pests and products (Fig. 2). The management–landscape continuum establishes the degree to which the natural system is altered to achieve each agroecosystem production goal. These changes in natural systems, primarily through perturbations in pattern variables, could influence the relative emphasis of process or pattern functions in consumer management (Rabb, 1978; Altieri and Whitcomb, 1980; Stinner et al., 1982). We refer to the organization hierarchy of Figure 1 for the appropriate fit of pattern and process factors to various management options.

INTENSIVELY MANAGED AGROECOSYSTEMS

Intensively managed agroecosystems are characterized by the production of genetically homogeneous populations through maximum use of resource inputs. This management approach focuses on process functions and results in decreased environmental heterogeneity (Fig. 2). Examples of U.S. intensive agricultural systems would include Iowa corn and Nebraska wheat regions, Florida and California vegetable-growing areas, Southern cotton production, New York apple orchards, Colorado feedlots, Georgia broiler production, and Illinois hog confinement systems. Maximizing producer and consumer yields in these systems depends on management of physiological factors, energy flux, and nutrient cycling. For example, one can model U.S. corn production by focusing on an individual plant or hectare and a series of machinery, fuel, plant breeding/physiology, nutrient, labor, and pesticide inputs (Pimentel et al., 1973). The homogeneous, static nature of the landscape pattern results in potential coarse–grained utilization of resources by the consumers. Thus, intensively managed agroecosystems are dominated by activities that occur within the process hierarchy (Fig. 1).

Row Crop Systems

The production of row crops results in substantial consumer pest problems. For example, an estimated 25% of all crops are lost annually to insects and pathogens in the United States (USDA, 1965). Management approaches that attempt to maximize primary productivity in these agroecosystems always include pest control strategies (Smith and Pimentel, 1978).

Chemical pesticides represent an important management tool for pest control (Eichers, 1981). Of the estimated 1 billion pounds of pesticides used in the United States, 49% is applied for insect and pathogen control (Pimentel et al., 1978; Berry, 1978). Approximately 85 million acres of row crops are treated annually with insecticides and fungicides (Eichers, 1981). The intensively managed agroecosystems of corn, cotton, soybeans, small grains, tobacco, peanuts, and fruits account for over 90% of the insecticide used in U.S. agriculture. The continuous production of row crop monocultures over the same land area often depends on chemical pest control. For example, chemical control of the corn rootworm is essential for continuous corn production in much of the U.S. Corn Belt (Luckmann, 1978). Thurston (1978) states that without insecticides potato yields would be severely reduced and some areas of the United States could no longer produce economic yields.

The physiological approach of chemical pest control and the ecosystem management strategy for row crop systems are linked within the process of hierarchy (Fig. 1). The materials are relatively cheap and robust in action against the consumer–producer trophic link (Muir, 1978). Since the goal of most row crop systems is to maximize the yield of a single-crop species over relatively large land areas, pest management based solely on the application of chemicals does not attempt to incorporate pattern in pest control.

Plant resistance to control consumer pests represents another process-oriented management tool for row crop production. Plant breeding for pest problems alters the genetic composition of crops and confers a degree of physiological resistance to insects and pathogens. Examples of the use of plant resistance to control consumer pests include the Hessian fly–wheat interaction (Hatchett and Gallun, 1970), control of stem rust and crown rust in oats (van der Plank, 1968), and the incorporation of resistance to a variety of insects and pathogens in potatoes (Thurston, 1978).

While the use of chemical pesticides and the development of plant resistance to insects and disease operates within the process hierarchy, the extensive geographic coverage of row crop systems necessitates the introduction of pattern in the management of row crop pests. Pimentel (1977) suggests that the origins of many row crop pest problems can be traced to management approaches that result in decreased environmental heterogeneity. The static, homogeneous landscape pattern that develops in these systems conflicts with many alternative methods of consumer pest control (Pimentel, 1961; Price and Waldbauer, 1975; Altieri and Whitcomb, 1980). The introduction of a uniform producer resource base may encourage pest outbreaks under a coarse-grained consumer exploitation strategy. Introduction of genetic and species diversity, spatial heterogeneity of physical variables, and temporal shifts in crop patterns offer a variety of alternatives to process-oriented control strategies (Stinner et al., 1982). Indeed, before World War II and the advent of modern chemical pest control and plant breeding, pattern approaches, such as crop rotations and selective burning, dominated pest management strategies (Muir, 1978).

Current ideas on pest control in row and orchard crops focus on integrated pest management (O'Brien, 1978). Integrated pest management strategies implicitly recognize the interconnections of the pattern and process hierarchy (Fig. 1). Problems related to consumer pest control reside at different levels in this hierarchy. For example, chemical pest control crosses the physiological and population levels of this hierarchy, while pest problems that arise as a function of crop monocultures occur at the community, landscape and ecosystem levels. Thus, a chemical "fix" will not completely solve a pest problem whose origins reside at the landscape level of organization. However, recognition of the connections between hierarchical levels encourages the use of integrated pest management strategies. Thus, corn pest problems can be best controlled through a combination of plant resistance, insecticides, and rotations with other crops.

Tree Plantations

Silviculture occupies a position somewhat analogous to agronomy in row crop systems in that it is concerned with the management of producers (trees) to enhance primary production (Smith, 1962). The natural forest landscape is altered through control of species composition and stand density. For purposes of our discussion we define tree plantation systems as the goal-oriented production of individual tree species. These agroecosystems range from Southern pine plantations through short-rotation, closely spaced tree populations for woody biomass production. Resource inputs are fewer for tree plantations than row crop systems (Fig. 2). These reduced inputs are primarily a function of the lower economic return for timber, pulp, and woody biomass systems relative to row crop systems (Hyde, 1980). This type of agroecosystem (except for certain experimental woody biomass plantations, USDA, 1980) has fewer process–level controls (e.g., pesticides and fertilization) to enhance primary productivity. Decreased management inputs result in increased spatial heterogeneity as a component of landscape pattern (Fig. 2). The growth form allows vertical structure to develop and the rotation length between planting and harvest increases plant community age and diversity. For example, loblolly pine plantations develop a significant understory of grasses, forbs, and shrubs (Chapman, 1942). Slash pine plantations in Florida develop a diverse herbaceous flora (Swindel et al., 1983). The fewer process-level controls and the increased spatial and temporal heterogeneity influence the management of consumers in these agroecosystems.

The management of consumer pests in tree plantations is focused primarily around the pattern hierarchy of populations, communities, and landscapes (Fig. 1). The relative lack of plant resistance and chemical controls in these systems necessitates management approaches addressed at pattern-oriented techniques (NAS, 1975). These involve selective cutting, maintenance of genetic diversity, and alteration of site conditions.

Management of Southern pine beetle, *Dendroctonus frontalis*, demonstrates pest control approaches in highly managed tree plantations. The insects attack almost any Southern pine species in large numbers, and an infected tree commonly dies within 3 weeks (Garan and Coster, 1968). As the infestation develops, scattered small groups of trees in dense stands are generally attacked (NAS, 1975). The infestation spreads rather rapidly through large areas of homogeneous pine populations during the latter stages of attack. In 1973 the infested area included almost 25 million hectares (Kucera and Barry, 1973). Control measures focus on prompt harvesting, improving stand characteristics through genetic diversity, thinning, removal of high-risk trees, avoidance of poor growing sites, and, for very high-valued trees, chemical spraying. Thus, the preferred strategy emphasizes pattern-related functions rather than physiological and ecosystem approaches to pest management.

An exception to the pattern-oriented approach involves woody biomass plantations that mimic row crop production in terms of the level of resource inputs and landscape changes. Indeed, Dwinell and Yates (1979) suggest the woody biomass plantations, such as short-rotation, closely spaced poplar systems, will have pest problems similar to row crop and orchard systems. Pest management strategies will then have to duplicate efforts now used in row crop and orchard systems.

Animal Confinement Systems

These agroecosystems, as a subset of animal grazing systems, rely on relatively high quality agricultural land to produce grains or other concentrated feed inputs for livestock (Krummel and Dritschilo, 1977). As products, consumers in these systems are removed from direct ecological connections with the producer system. For example, while the number of beef cattle in U.S. feedlots can influence the economics of corn production, increased corn productivity does not require management or knowledge of the beef production system. This is also true for U.S. broiler, egg, hog, and dairy confinement systems (Pimentel et al., 1975).

Animal husbandry techniques rely on physiological parameters that increase meat, egg, and milk production. In addition, machinery and energy inputs enhance management options to improve animal yields (Pimentel et al., 1975). For example, changes in diet and improved breeding programs have increased the efficiency of broiler production; broilers convert about 2 kg of feed into 1 kg of liveweight (Benson and Witzig, 1977). By removing these animal systems from direct ecological links with producers, management systems can concentrate entirely on process functions and ignore problems that are related to changes in landscape pattern.

NON-INTENSIVE MANAGEMENT LEVELS

Non-intensively managed agroecosystems are characterized by the goal-oriented production of plant and animal populations through limited use of management inputs (Fig. 2). The low economic return per land area limits resource inputs to a greater degree than in the intensively managed agroecosystems (Hyde, 1980; Pimentel et al., 1980). Examples of these systems include rangeland grazing and harvests of timber in national forestland. Limited resource inputs focus management approaches at the landscape, community, and ecosystem levels of the ecological hierarchy (Fig. 1). It is at these hierarchical levels that the link between process and pattern will be most evident.

Animal Grazing Systems

These management systems are quite varied and are mainly a function of the types of large herbivores and the specific regions of the United States (Ward et al., 1977; Pimentel et al., 1980). The management–landscape continuum ranges from managed systems to those with very few resource inputs. Landscape pattern ranges from relatively static and homogeneous (e.g., irrigated pasture) to dynamic and heterogeneous landscape areas (e.g., semiarid to arid rangeland) where little management is practiced (Fig. 2). Coarse-grained or fine-grained consumer exploitation of the producer resource base can occur in animal grazing systems. Coarse-grained utilization of the resource base would develop in managed pastures. In this situation producer diversity is reduced and consumers must concentrate on a few populations. In situations such as semiarid rangeland, lower management levels result in greater producer diversity. Thus, consumers can utilize the producer base in a fine-grained manner.

While it is true that climatic and physiographic variables influence the development of natural landscape patterns, consumers in grazing lands have a prominent place and function in shaping this pattern. Here it is necessary to focus on the role of process and pattern in our hierarchy of ecological systems (Fig. 1). Indeed, with the diversity of management practices and the spatial area that animal grazing systems occupy, producer–consumer interactions may involve all levels of organization in this hierarchy.

Herbivores can influence spatial heterogeneity in the producer complex. In pristine grazing lands ungulate feeding behavior and regional climatic variables tend to govern vegetation patterns (McNaughton, 1976; Coppock et al., 1983a,b). In modern managed systems land allocation based on ownership or management responsibility also determines vegetation patterns. For example, summer–winter grazing regimes and stocking densities in the West tend to set the plant community of each particular area (Van Dyne et al., 1980). The shift from midgrass prairie to shortgrass prairie occurs with the advent of continuous grazing pressure from cattle (Lewis, 1971). However, the development of plant species composition is not well understood, and there is much conflicting information (Rice and Westoby, 1978; Van Poolen and Lacey, 1979; Lacey and Van Poolen, 1981).

In addition to altering species composition, animal grazing systems affect the pattern of various biogeochemical and physical parameters. Woodmansee (1978) considers consumers, especially large herbivores, to be the single largest factor in determining system nitrogen losses and important vectors for nutrient transport from one area to another. Overstocking of grazers can affect physical parameters on the landscape by eliminating vegetative cover that results in severe erosion on steep ground. However, it has been argued that the presence of an optimal level of large grazers is necessary to maintain high-quality rangeland (Reardon and Merrill, 1976).

Process in rangeland management and animal grazing systems operates through physiological and ecosystem links between the producers and consumers (Fig. 1). For example, many plants are sensitive to very slight grazing pressure (Chew and Rodman, 1979). In areas that are relatively devoid of herbivore pressure, these grazing-sensitive plants will be found. On the other hand it is known that some plants develop and thrive under grazing pressure (Pitt and Heady, 1979; Dyer et al., 1982; Reardon and Merrill, 1976). To organize physiological processes and ecosystem productivity as influenced by these grazing pressures, a herbivore optimization process has been introduced (McNaughton, 1976, 1979; Dyer et al., 1982). This hypothesis simply states that grassland conditions, principally productivity measures, will at first increase with grazing pressure, after which there will be a decrease.

With the complex array of animal grazing systems, it is not surprising the literature dwells not only on ecosystem processes and physiological events but also on plant community composition and landscape interactions. Within the process and pattern hierarchy and the management–landscape continuum, one needs to carefully examine each problem of grazing management to determine in which level of organization in the ecological hierarchy the problem may reside. This will help determine whether process or pattern factors or a combination of both will be appropriate for a management strategy. Also, it is important to regard all consumers in grazing lands. Large herbivores constitute only a small percentage of the animals present in the system (French et al., 1979). These animals will also influence process and pattern functions in these ecological systems.

Systems Containing National Forest, Wildlife Refuges, and Other Multiple-Use Lands

Along our continuum (Fig. 2) lie quasi-natural systems (i.e., national forests) that are subjected to varying degrees of management, often for a main producer or consumer product but with secondary products to satisfy a variety of societal needs (Krutilla and Fisher, 1975). The cutting of timber patches, creation of impoundments by damming flowing streams, development of campgrounds, roads, and trails, and creation of grazing allotments create new landscape patterns. These new patterns, many of which are specific to a particular use category, often lie side-by-side. Their impacts on the landscape can subsequently create conflicting interactions among consumers that ordinarily would not exist.

One major facet controlling landscape pattern in multiple-use systems is the overall size of the area being considered. A second is the type of management goal, particularly for consumers (including humans involved in recreational activities), and a third is location. In some areas multiple-use practices produce little effect on natural pattern because the area is so large that it can be indistinguishable from wilderness. However, in areas close to population centers that combine restricted size with several use potentials, consumer management takes place within a complex array of natural and managed landscape patterns. Since it is not our purpose to dwell on all possibilities, we will not examine the issue further, but we refer the reader to two interesting and important books dealing with some of these aspects (Clawson and Knetsch, 1966; Krutilla and Fisher, 1975).

Wilderness Systems

At first glance it might seem out of context to regard wilderness systems in a chapter that deals with the role of consumers in agroecosystems. However, we use wilderness systems to formulate the basis for our management–landscape continuum (Fig. 2). Indeed, all managed agroecosystems were once wilderness systems before the perturbations of goal-oriented production systems were introduced. Except for wildfire control, most wilderness systems are considered unchanged, or at the very least, little changed from the standpoint of the historical biotic and abiotic forces that formed these landscapes. Thus, the processes and patterns that make up these natural systems form the foundations of the various levels of organization that occur in the hierarchy. Changes in these systems (i.e., the introduction of cattle grazing or timber harvesting) would alter landscape pattern. If this is indeed the case, then it is of heuristic value to understand current trophic interactions that contribute to the maintenance of these systems. This value constitutes a major argument for the continuance of undisturbed areas.

SUMMARY

Structure and design in landscape pattern provide a focal point for examining the link between agricultural management practices and consumer/producer interactions. Consumers, as products or pests, develop different modes of interaction with producers as a function of the effect of management intensity on landscape pattern. However, to understand events at the landscape level it is necessary to compare this level of organization with more traditional levels, such as populations, communities, or ecosystems. We view landscapes as occupying a position in a hierarchy of ecological organization in which pattern analyses are the basic mode of identifying relationships. Process analyses, as a complement to pattern, examine functional relationships among the biotic and abiotic components of the natural system. Emphasis can be placed on physiology, energy flow, and nutrient cycling. This approach provides information about the dynamics of the organism, ecosystem, or biosphere. By integrating pattern and process, one can effectively match an environmental problem to an appropriate hierarchical level of the ecological system. This is especially helpful in managing consumers in agroecosystems.

To explore these ideas we presented a continuum of agroecosystem types, ranging from row crops to wilderness. The continuum examined the effect of management intensity on spatial heterogeneity as a component of landscape pattern. High management inputs result in static, homogeneous landscape patterns, while few management inputs lead to relatively more dynamic, heterogeneous patterns. We then examined consumers, as pests and products, in the various agricultural systems that exist along this management continuum of landscape interactions. Consumers both influence and are influenced by the pattern that develops in these agroecosystems. In highly managed systems in which landscape pattern is relatively homogeneous, the management of consumers relies on process factors. These range from chemical control of pests through the physiological changes in animal breeding strategies. Less managed systems result in more dynamic patterns, and consumer management focuses on pattern factors. These range from a variety of large herbivore grazing strategies on western range through selective cutting in southern pine forests. It is for this reason that we urge that all consumers be considered when management practices may alter the natural landscape pattern.

REFERENCES

Allen, T. F. H. and Starr, T. B. (1982). *Hierarchy: Perspectives for Ecological Complexity.* University of Chicago Press, Chicago.

Altieri, M. A. and Whitcomb, W. H. (1980). Weed manipulation for insect pest management in corn. *Environ. Manage.* **4:**483–489.

Beals, E. W. (1969). Vegetational change along altitudinal gradients. *Science* **165:**981–985.

Benson, V. W. and Witzig, T. J. (1977). The Chicken Broiler Industry: Structure, Practices, and Costs. Agri. Econ. Rep. No. 381, Econ. Res. Ser., U.S. Department of Agriculture, Washington, D.C.

Berry, J. H. (1978). Pesticides and energy utilization, paper presented at AAAS annual meeting in Washington, D.C.

Burgess, R. L. and Sharpe, D. M., (eds.) (1981). *Forest Island Dynamics in Man-Dominated Landscapes.* Springer-Verlag, New York.

Chapman, H. H. (1942). Management of loblolly pine in the pine-hardwood region in Arkansas and in Louisiana west of the Mississippi River, Yale University School of Forestry, Bulletin 49.

Chew, F. S. and Rodman, J. E. (1979). Plant resources for chemical defense. In *Herbivores: Their Interaction with Secondary Plant Metabolites.* G. A. Rosenthal and D. H. Janzen (eds.), Academic Press, New York, pp. 271–307.

Clark, W. C., Jones, D. D., and Holling, C. S. (1978). Patches, movements, and population dynamics in ecological systems: A terrestrial perspective. In *Spatial Pattern in Plankton Communities*. J. A. Steele (ed.), Plenum Press, New York.

Clawson, M. and Knetsch, J. L. (1966). *Economics of Outdoor Recreation*. The Johns Hopkins University Press, Baltimore.

Cohen, J. E. (1970). A Markov contingency table model for replicated Lotka-Volterra systems near equilibrium. *Am. Nat.* **104:**547–559.

Coppock, D. L., Detling, J. K., Ellis, J. E., and Dyer, M. I. (1983a). Plant-herbivore interactions in a North American mixed-grass prairie. I. Effects of black-tailed prairie dogs on intraseasonal aboveground plant biomass and nutrient dynamics and plant species diversity. *Oecologia*, in press.

Coppock, D. L., Ellis, J. E., Detling, J. K., and Dyer, M. I. (1983b). Plant-herbivore interactions in a North American mixed-grass prairie. II. Responses of bison to modification of vegetation by prairie dogs. *Oecologia*, in press.

Drury, W. H. and Nisbet, I. C. T. (1973). Succession. *J. Arnold Arbor. Harvard Univ.* **54:**331–368.

Dyer, M. I., Detling, J. K., Coleman, D. C., and Hilbert, D. W. (1982). The role of herbivores in grasslands. In *Grasses and Grasslands: Systematics and Ecology*. J. R. Estes, R. J. Tyrl, and J. N. Brunken (eds.), University of Oklahoma Press, Norman, pp. 255–295.

Dwinell, L. D. and Yates, H. O. (1979). Potential entomological and pathological problems of forest tree crops grown under intensive culture, Southeastern Forest Experiment Station, Athens, Georgia.

Ehrlich, P. R. and Raven, P. H. (1969). Differentiation of population, *Science* **165:**1228–1232.

Eichers, T. R. (1981). Use of pesticides by farmers. In *Handbook of Pest Management in Agriculture*. Vol. II. D. Pimentel (ed.), CRC Press, Boca Raton, Florida, pp. 3–25.

French, N. R., Steinhorst, R. K., and Swift, D. M. (1979). Grassland biomass trophic pyramids. In *Perspectives in Grassland Ecology*. N. R. French (ed.), Ecological Studies 32, Springer-Verlag, New York, pp. 59–87.

Forman, R. T. T. and Godron, M. (1981). Patches and structural components for a landscape ecology. *Bioscience* **31:**733–740.

Garan, R. I. and Coster, J. E. (1968). Studies on the attack behavior of the Southern pine beetle. III. Sequence of tree infestations within stands. *Contrib. Boyce Thompson Inst.* **24:**77–86.

Goh, B.-S. (1980). *Management and Analysis of Biological Populations*. Elsevier Scientific Publishing, Amsterdam.

Hatchett, J. H. and Gallun, R. L. (1970). Genetics of the ability of the Hessian fly, *Mayetiola destructor*, to survive on wheats having different genes for resistance. *Ann. Ent. Soc. Am.* **63:**1400–1407.

Holdridge, L. (1967). *Life Zone Ecology*. Tropical Science Center. San Jose, Costa Rica.

Holling, C. S. (1966). The functional response of invertebrate predators to prey density. *Mem. Entomol. Soc. Con.* **48:**1–85.

Horn, H. S. and MacArthur, R. H. (1972). Competition among fugitive species in a harlequin environment. *Ecology* **53:**749–752.

Huffaker, C. B. (1958). Experimental studies on predation: dispersion factors and predator-prey oscillations. *Hilgardia* **27:**343–383.

Hyde, W. F. (1980). *Timber Supply, Land Allocation, and Economic Efficiency.* The Johns Hopkins University Press, Baltimore.

Krummel, J. R. and Dritschilo, W. (1977). Resource costs of animal protein production. *World Animal Rev.* **21:**6–10.

Krummel, J. R. and Hough, J. (1980). Pesticides and controversies: Benefits versus costs. In *Pest Control: Culture and Environmental Aspects.* D. Pimentel and J. H. Perkins (eds.), Westview Press, Boulder, Colorado, pp. 159–179.

Krummel, J. R., Gilmore, C. C. and O'Neill, R. V. (1983). Siting field research in the western Kentucky energy development region: an exploration of a regional approach, ORNL/TM-8581. Oak Ridge National Laboratory, Oak Ridge, Tennessee.

Krutilla, J. V. and Fisher, A. C. (1975). *The Economics of Natural Environments.* The Johns Hopkins University Press, Baltimore.

Kucera, D. R. and Barry, P. J. (1973). Southern pine beetle at epidemic proportions. *For. Farm.* **33:**16–17, 34.

Lacey, J. R. and Van Poolen, H. W. (1981). Comparison of herbage production on moderately grazed and ungrazed western ranges. *J. Range Manage.* **34:**210–212.

Levin, S. A. (1976). Spatial patterning and the structure of ecological communities. *Lecture on Mathematics in the Life Sciences* **8:**1–35.

Lewis, J. K. (1971). A synthesis of structure and function, 1970. In *Preliminary Analysis of Structure and Function in Grasslands.* N. R. French (ed.), Range Science Department, Science Series 10, Colorado State University, Fort Collins, pp. 317–387.

Luckmann, W. H. (1978). Insect control in corn-practices and prospects. In *Pest Control Strategies.* E. H. Smith and D. Pimentel (eds.), Academic Press, New York, pp. 137–156.

MacArthur, R. H. and Levins, R. (1964). Competition, habitat selection and character displacement in a patchy environment, *Proc. Natl. Acad. Sci.* **51:**1207–1210.

May, R. M. (1975). Patterns of species abundance and diversity. In *Ecology and the Evolution of Communities.* M. L. Cody and J. M. Diamond (eds.), Belknap Press, Cambridge, Massachusetts, pp. 81–120.

McNaughton, S. J. (1976). Serengeti migratory wildebeest: Facilitation of energy flow by grazing. *Science* **91:**92–94.

McNaughton, S. J. (1979). Grazing as an optimization process: Grass-ungulate relationships in the Serengeti. *Am. Nat.* **113:**691–703.

McNaughton, S. J. and Wolf, L. L. (1979). *General Ecology.* Holt, Rhinehart and Winston, New York.

Muir, W. (1978). Pest control—a perspective. In *Pest Control Strategies.* E. H. Smith and D. Pimentel (eds.), Academic Press, New York, pp. 3–8.

NAS. (1975). *Pest Control: An Assessment of Present and Alternative Technologies.* Vol. IV, *Forest Pest Control.* National Academy of Sciences, Washington, D.C.

O'Brien, R. D. (1978). Integrated pest management—a biological viewpoint. In *Pest Control Strategies.* E. H. Smith and D. Pimentel (eds.), Academic Press, New York, pp. 23–40.

Odum, E. P. (1971). *Fundamentals of Ecology*. 3rd ed. Saunders, Philadelphia.

Paine, R. T. (1966). Food web complexity and species diversity. *Am. Nat.* **100**:65–75.

Pielou, E. C. (1977). *Mathematical Ecology*. Wiley-Interscience, New York.

Pimentel, D. (1961). Species diversity and insect population outbreaks. *Ann. Entomol. Soc. Am.* **54**:76–86.

Pimentel, D. (1977). The ecological basis of insect pest, pathogen and weed problems. In *Origins of Pest, Parasite, Disease and Weed Problems*. J. M. Cherrett and G. R. Sagar (eds.), Blackwell Scientific Publications, Oxford, pp. 1–33.

Pimentel, D., Hurd, L. E., Bellotti, A. C., Forster, M. J., Ika, I. N., Sholes, O. D., and Whitman, R. J. (1973). Food production and the energy crisis. *Science* **182**:443–449.

Pimentel, D., Dristschilo, W., Krummel, J. R., and Kutzman, J. (1975). Energy and land constraints in food protein production. *Science* **190**:754–761.

Pimentel, D., Krummel, J., Gallahan, D., Hough, J., Merrill, A., Shreiner, I., Vittum, P., Koziol, F., Back, E., Yen, D., and Fiance, S. (1978). Benefits and costs of pesticides in U.S. food production. *Bioscience* **28**:772, 778–784.

Pimentel, D., Oltenacu, P. A., Nescheim, M. C., Krummel, J., Allen, M. S., and Chick, S. (1980). The potential for grass-fed livestock: resource constraints. *Science* **207**:843–848.

Pimm, S. L. (1982). *Food Webs*. Chapman-Hall, London.

Pitt, M. D. and Heady, H. F. (1979). The effects of grazing intensity on annual vegetation. *J. Range Manage.* **32**:109–114.

Price, P. W. and Waldbauer, G. P. (1975). Ecological aspects of pest management. In *Introduction to Insect Pest Management*. R. L. Metcalf and W. H. Luckman (eds.), Wiley-Interscience, New York, pp. 37–73.

Rabb, R. L. (1978). A sharp focus on insect populations and pest management from a wide-area view. *Bull. Entomol. Soc. Am.* **24**:55–61.

Reardon, P. O. and Merrill, L. B. (1976). Vegetation response under various grazing management systems in the Edwards Plateau of Texas. *J. Range Manage.* **29**:195–198.

Rice, B. and Westoby, M. (1978). Vegetative responses of some great basin shrub communities protected against jackrabbits or domestic stock. *J. Range Manage.* **31**:28–34.

Ricklefs, R. E. (1979). *Ecology*. 2nd ed., Chiron Press, Portland, Oregon.

Root, R. B. (1973). Organization of a plant-arthropod association in simple and diverse habitats: the fauna of collards (*Brassica oleracea*). *Ecol. Monogr.* **43**:9–124.

Shugart, H. H. and West, D. C. (1977). Development of an Appalachian deciduous forest succession model and its application to assessment of the impact of chestnut blight. *J. Environ. Manage.* **5**:161–179.

Shugart, H. H. and O'Neill, R. V. (eds.) (1979). *Systems Ecology*. Dowden, Hutchinson and Ross, Stroudsburg, Pennsylvania.

Smith, D. M. (1962). *The Practice of Silviculture*. John Wiley & Sons, New York.

Smith, E. H. and Pimentel, D., (eds.) (1978). *Pest Control Strategies*. Academic Press, New York.

Spedding, C. R. W. (1975). *The Biology of Agricultural Systems*. Academic Press, New York.

Steele, J. H. (1974). *The Structure of Marine Ecosystems*. Harvard University Press, Cambridge, Massachusetts.

Stinner, R. E., Regniere, J., and Wilson, K. (1982). Differential effects of agroecosystem structure on dynamics of three soybean herbivores. *Environ. Entomol.* **11**:538–534.

Swindel, B. F., Conde, L. F., and Smith, J. E. (1983). Plant cover and biomass response to clear-cutting, site preparation, and planting in *Pinus eliottii* flatwoods. *Science* **219**:1421–1422.

Thurston, H. D. (1978). Potentialities of pest control in potatoes. In *Pest Control Strategies*. E. H. Smith and D. Pimentel (eds.), Academic Press, New York, pp. 117–136.

USDA. (1965). *Losses in Agriculture*. Agr. Handbook No. 291. Agri. Res. Serv. U.S. Government Printing Office, Washington, D.C.

USDA. (1980). Energy and wood from intensively cultured plantations: research and development program. U.S. Forest Service, U.S. Department of Agriculture, General Technical Report NC-58. North Central Experiment Station.

U.S. Department of Agriculture (USDA). 1982. Agricultural Statistics 1982. U.S. Department of Agriculture, Washington, D.C.

Van der Plank, J. E. (1968). *Disease Resistance in Plants*. Academic Press, New York.

Van Dyne, G. M. (1966). Ecosystems, Systems Ecology, and System Ecologists, ORNL-3957. Oak Ridge National Laboratory, Oak Ridge, Tennessee.

Van Dyne, G. M., Brockington, N. R., Szocs, Z., Duek, J., and Ribic, C. A. (1980). Large herbivore subsystem. In *Grasslands, Systems Analysis and Man*. A. J. Breymeyer and G. M. Van Dyne (eds.), International Biological Programme 19, Cambridge University Press, Cambridge, pp. 269–537.

Van Poolen, H. W. and Lacey, J. R. (1979). Herbage response to grazing systems and stocking intensities. *J. Range Manage.* **32**:250–253.

Ward, G. M., Knox, P. L., and Hobson, B. W. (1977). Beef production options and requirements for fossil fuel. *Science* **198**:265–271.

Whittaker, R. H. (1953). A consideration of climax theory: the climax as population and pattern. *Ecol. Manage.* **23**:41–78.

Whittaker, R. H. (1975). *Communities and Ecosystems*. 2nd ed. MacMillan, New York.

Wilson, E. O. and Willis, E. O. (1975). Applied biogeography. In *Ecology and Evolution of Communities*. M. L. Cody and J. M. Diamond (eds.), Harvard University Press, Cambridge, Massachusetts, pp. 522–534.

Woodmansee, R. G. (1978). Additions and losses of nitrogen in grassland ecosystems. *Bioscience* **28**:448–453.

The Positive Interactions in Agroecosystems

D.A. Crossley, Jr. and Garfield J. House

Institute of Ecology
The University of Georgia
Athens, Georgia

Renate M. Snider and Richard J. Snider

Department of Zoology
Michigan State University
East Lansing, Michigan

Benjamin R. Stinner

Department of Entomology
Ohio State University,
Ohio Agricultural Research Center
Wooster, Ohio

It has been 30 years since Eugene Odum rescued the term *ecosystem* from relative obscurity and set in motion the modern study of ecological systems. In his book *Fundamentals of Ecology* (1953) Odum used the ecosystem concept as an organizing principle, in contrast to the autecological-ecophysiological approaches prevalent at the time. The ecosystem approach gave scientists guiding principles through which environmental problems could be effectively addressed. Autecology does not provide the tools needed to deal with large-scale environmental issues, be they radioactive fallout or abuse of agricultural chemicals. As Odum emphasized, an ecosystem is a functional entity that is more than the sum of its parts. Knowing the properties of the parts does not adequately predict the behavior of the whole.

Odum's holistic approach was timely; since the late 1950s, awareness of rapidly deteriorating environmental quality had been increasing; urgently needed decisions affecting whole systems had to be made. With the coming of age of ecosystem ecology in the 1960s, regulations governing chemical releases and usages could be developed. Their impact was especially pronounced in agriculture. Today, the importance and usefulness of the holistic approach and the urgency of preventing further environmental deterioration are still pertinent for modern agriculture.

The term *ecosystem* has been used in a variety of ways; the term *agroecosystem*, a result of applying ecosystem concepts to agriculture, has been particularly variable. Like other authors, we feel obliged to discuss it and to define how we use it here. All ecosystems on our planet are more or less open, exchanging abiotic and biotic elements with other ecosystems. Agroecosystems are designed to be extremely open, with major exports of primary or secondary production. Modern agroecosystems are entirely dependent on human intervention; without management agroecosystems as such will not persist. For this reason they are often designated as artificial ecosystems as opposed to natural systems, which do not require man's management. In fact, the great majority of our so-called natural systems— forests, rivers, lakes—are managed so that the distinction between natural and artificial is no longer clear. Definitions of agroecosystems often include an entire support base of energy and material subsidies, seeds and chemicals, and even a sociopolitico–economic matrix in which management decisions are rooted. That is no doubt logical, but we prefer to designate the individual field as the agroecosystem because this is consistent with designating and studying an individual forest catchment or a lake as an ecosystem. We envision the *farm system* as consisting of a set of agroecosystems— fields with similar or different crops—together with the support mechanisms and socioeconomic factors contributing to their management. Use of individual field units as agroecosystems is the approach we have taken in studies at the University of Georgia. There we are applying methods of ecosystem sampling and analysis in order to answer an initial, essential question: Are agroecosystems mere collections of components managed by man, or do they retain the functional properties of natural ecosystems—nutrient conservation mechanisms, energy storage and utilization patterns, and regulation of biotic diversity? The answer seems to be in the affirmative, at least for the systems in Georgia (Stinner et al., in press).

Analysis of ecosystems has centered on measurements of three general types of phenomena: species richness in various components; functional properties such as energy flow and nutrient cycling; and more recently, ecosystem response to perturbation (Botkin, 1980). A fourth phenomenon, likely to be important, is allelochemical regulation (Rice and Pancholy, 1973). In this analysis we will focus on the positive interactions among components of ecosystems. By positive interactions we mean those that generally lead to the maintenance of functional integrity, biotic persistence, and self-regulation. In contrast, negative interactions would be those that lead to deterioration of biotic persistence and loss of functional integrity. In order to promote efficient resource use coupled with persistence of the systems through time, agroecosystem management of the future will attempt to benefit from such positive interactions. We equate positive interactions with cooperative relationships among organisms, where interaction is mutually beneficial. Some modern agronomists perceive noncrop organisms as exclusively antagonistic. We suggest that the designations between antagonistic and cooperative interactions exist in a continuum, and that management practices may determine the prevalence of one or the other.

Figure 1 contrasts the relationships between components for natural and agricultural ecosystems. Figure 1A illustrates nutrient flow between major components of a natural ecosystem. Figure 1B shows how nutrient flows are changed by management activities. Transfer from primary producers to consumers is reduced by applications of insecticides (shown as a control element in Fig. 1B). Transfer from primary producers to the decomposition subsystem is regulated by herbicides and cultivation during the growing season, changing the timing of inputs to the decomposers. At season's end the primary producers may be mowed down, providing massive pulse to the decomposers. The decomposition subsystem itself is inadvertently regulated both by application of pesticides and by cultivation. Plowing incorporates crop residues into the mineral soil, speeding their decomposition rates.

The overall impact of these regulatory measures is to upset the normally conservative nutrient cycles by increasing rates of nutrient release to the abiotic phase. In Figure 1B we show a relatively large nutrient influx (fertilizer) and export of primary production. As a consequence of high influx and disturbance of ecosystem function, a relatively large nutrient loss (runoff or "nonpoint pollution") is suggested.

The four-compartment model of Figure 1 is an oversimplified abstraction of ecosystem nutrient cycling, but it does serve to illustrate the negative impact of some management technologies. Self-regulatory properties, prominent in natural systems, are altered or suppressed in agricultural systems. Within-system interactions may not differ in kind, but certainly differ in degree. In the following we will briefly consider some of the positive interac-

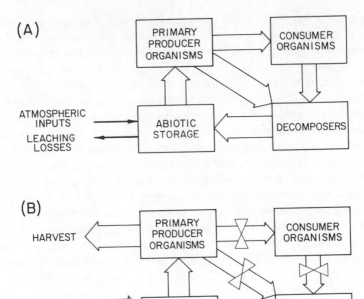

Figure 1. Nutrient circulation through major components of (A) natural ecosystems and (B) agroecosystems. In natural systems internal cycling of nutrients greatly exceeds flowthrough expressed as atmospheric inputs and leaching losses (top). In agroecosystems management techniques include control of consumer organisms with insecticides, control of primary producers with herbicides as well as cultivation and harvest, and control of decomposition through cultivation and plowing. Fertilizer becomes the major nutrient input to the agroecosystem, and harvest constitutes a major output, but leaching losses become larger also.

tions characteristic of agroecosystems, that is, those which tend to increase self-regulation of processes. We suggest that agroecosystem management of the future will attempt to incorporate such positive interactions into modified management practices.

PRIMARY PRODUCERS

In field crop agroecosystems, crop monoculture is a goal seldom achieved. Weeds are generally viewed as competitors for water, nutrients, or sunlight, or at least viewed as nuisances. Ecologists have observed that weed species often have some predictable biological characteristics such as resistance to climatic extremes, high reproductive and dispersal capacities, and allelopathic chemicals. The autecology of weed species is a rapidly developing research area (Harper, 1960). Since weed-free fields are unusual, it follows that most crop plants can tolerate weeds at low densities without dramatic yield reductions. Are there, in fact, positive interactions—mutualistic associations—between weeds and crop plants? A few tentative suggestions have been made (Sagar, 1968; Tripathi, 1977). Elimination of present-day weeds might open a niche available to even better competitors. Present-day weeds and crop plants may have evolved a mutual tolerance, if not a true mutualism (Maun, 1974). Elimination of weeds might increase certain pest problems (see below) and can contribute significantly to erosion problems, especially in the early growth and post-harvest period. These suggestions are largely speculative. It does appear that weeds constitute a significant nutrient reservoir in agroecosystems. Where competition for nutrients is not limiting, weedy growth can retain nutrients that might otherwise be lost from the agroecosystem (Stinner et al., in press). In some cases weed species may promote nitrogen fixation. The value of weeds for manipulation of beneficial insect populations (Altieri and Whitcomb, 1979) is increasingly suggested.

A philosophy of beneficial weed control could evolve in a manner similar to that of insect control. That is, the treatment of insect problems evolved from a strategy of complete insect control to a perspective of tolerance of some pests in low numbers. Economic and environmental considerations may modify the popular attitude that total weed control through chemical application is essential. Concurrently, management practices could progress toward techniques that encourage certain weed species, once beneficial interactions between crops and weeds become understood (Altieri and Whitcomb, 1979). Interplanting practices furnish an example of what may be regarded as progress in this direction; once considered too troublesome to use in large monocultures, the practice is currently gaining in popularity (Trenbath, 1981).

CONSUMERS

Because of their potential economic importance, insect herbivores in crop systems, and their responses to management practices, have been ecologically well defined. Except for pollinators, insect herbivores as a group are generally viewed as undesirable; the insect-free field, however, has been virtually impossible to obtain. Below, we outline interactions within the consumer community itself; specific insect–plant interactions will be considered in a subsequent section. Competition between herbivore insect species is often considered a major factor for shaping the structure of their community, but the magnitude of its effects has been overemphasized. Even in agricultural fields, primary consumption is seldom so high as to suggest food as a limiting resource. On a global scale, for instance, crop losses to insects before harvest have been estimated at 14% (Perkins, 1982).

Chemical control measures aimed at one or a few consumer species are not the sole determinants of species composition of the community. Barrett (1968) found that diversity was increased by a single insecticide application. Even after repeated attempts at chemical control, faunal diversity in crop fields may be high. Yet the significance of this diversity is largely unknown. For example, do these diverse arthropod communities include the potential for maintaining pest insect species below economic thresholds? It is probable that insect populations in a given agroecosystem are more strongly influenced by characteristics of surrounding ecosystems than by field-specific factors. Stinner et al. (1982) point out how vagile many pest species are and stress the importance of the matrix of crop fields and regional agricultural practices as determinants of community species composition. Such extrinsic factors may favor diversity not only of the consumers but may also play a role in maintaining predator–prey complexes. Integrated pest management (IPM) takes advantage of these to hold primary consumption to acceptable levels. While IPM uses a combination of means, TPM, or total pest management (Perkins, 1982), embodies the notion of complete eradication of pest species. As Perkins (1982) points out, both approaches seized upon ecological principles and the term *integrated* to characterize their efforts.

PRODUCER-CONSUMER INTERACTIONS

Insect–plant coevolution is a major unifying theme in insect ecology today. Often, it is viewed as a set of antagonistic exchanges in chemical warfare. Evidence for mutualism between plants and insects—positive interactions—is becoming more convincing (Owen, 1980; Owen and Wiegert, 1976; Petelle, 1982; Thompson, 1982). In agricultural systems the term *beneficials* is used to denote predators and parasites that may reduce pest insect populations. Altieri and Whitcomb (1979) have presented the argument that weeds are important in the biology of many of the beneficial insects. They suggest field management techniques to take advantage of this relationship and maximize numbers of beneficial insects. Aside from pollination, there have been few demonstrations of the value of herbivory to crop plants. However, Hilbert et al. (1981), McNaughton (1979), Dyer (1975), and others have argued that a low level of grazing may increase net primary production—that there is an optimal level of grazing in ecosystems. Evidence cited includes increased photosynthetic rates following grazing, reallocation of photosynthate, increased tillering or other elaboration, better light penetration, and a variety of indirect mechanisms.

Usually, any consumption of crop plants is considered to be antagonistic, even at relatively low levels. However, some consumption may be beneficial to crops. Compensatory growth following herbivory is well documented (Hills and Peters, 1971; Barner and Fletcher, 1974; Crossett et al., 1975). In field crop systems the exact nature and extent of plant–insect mutualism remains to be explored.

DECOMPOSITION AND SOIL

Soil—its structure, nutrients, and organic matter—is the major determinant of primary production in many regions. Yet, the ecology of the soil—the belowground ecosystem—is still relatively unknown.

From the viewpoint of nutrient dynamics in agroecosystems, soil is the major nutrient storage reservoir. As in other ecosystems, the complex interactions between roots, microbes, and animals contain both positive and negative elements (Curl, 1982), with the soil fauna acting as regulators rather than active decomposers. The burgeoning literature on mycorrhizal fungi indicates that plant–fungus mutualisms may be of considerable importance for agricultural crops. Dissemination of mycorrhizal spores by arthropods points to the importance of both mutualistic associations and indirect regulation. There is some evidence that soil arthropods may play a significant role in dispersing mycorrhizal propagules (Rabatin and Rhodes, 1982).

In agroecosystems chemical and mechanical disturbances influence both biotic and abiotic components of the decomposition subsystem, and their effects are superimposed on normal faunal and floral phenological cycles. The impact of various tillage practices on soil biota are not always clear-cut, however (Loring et al., 1981; Moore et al., in press). Research in no-tillage systems may hold the greatest promise for identifying positive interactions within the soil. In general, no-tillage provides a more favorable environment for soil–litter biota: crop residues are left on the soil surface, reducing moisture loss and providing relatively continuous substrate for decomposers; organic matter input is more gradual, as is nutrient release; faunal interactions are probably more pronounced and more significant in regulating decomposition rates under no-tillage conditions. Root growth, in turn, can be enhanced by increased invertebrate animal activity (Edwards and Lofty, 1977).

Weeds and weed residues, suppressed by conventional tillage practices, act as temporary nutrient reservoirs in no-tillage systems. Residue management research should uncover interactions between crops and weeds that are beneficial to crop-nutrient dynamics, but many of the intermediate coupling processes remain to be quantified.

This brief characterization of the positive interactions in agroecosystems suggests that mutualistic relationships are extensive, even in field crop agroecosystems. We predict that the identification of positive interactions in agroecosystems will continue to grow. These mutualistic interactions are characteristic of ecosystems with self-regulating capacities. They indicate that the term ecosystem is quite appropriate for field crop systems and that agroecosystems differ in degree, but not in kind, from other ecosystems.

REFERENCES

Altieri, M. A. and Whitcomb, W. H. (1979). The potential use of weeds in the manipulation of beneficial insects. *HortSci.* **14**:12–18.

Barner, R. and Fletcher, K. E. (1974). Insect infestations and their effects on the growth and yield of field crops: a review. *Bull. Entomol. Res.* **64**:141–160.

Barrett, G. W. (1968). The effects of an acute insecticide stress on a semi-enclosed grassland ecosystem. *Ecology* **49**:1019–1035.

Botkin, D. B. (1980). A grandfather clock down the staircase: Stability and disturbance in natural ecosystems. In *Forests: Fresh Perspectives from Ecosystem Analysis*, Richard H. Waring (ed.), Oregon State University Press, Corvallis, pp. 1–10.

Crossett, R. N., Campbell, D.J., and Stewart, H. E. (1975). Compensatory growth in cereal root systems. *Plant and Soil* **42**:673–683.

Curl, E. A. (1982). The Rhizosphere: Relation to Pathogen behavior and root disease. *Plant Disease* **66**:624–630.

Dyer, M. I. (1975). The effects of red-winged blackbirds (*Agelaius-phoeniceus*) on biomass production of corn grains (*Zea mays* L.). *J. Appl. Ecol.* **12**:719–726.

Ewards, C. A. and Lofty, J. R. (1977). The influence of invertebrates on root growth of crops with minimal or zero cultivation. *Ecol. Bull. (Stockholm)* **25**:348–356.

Harper, J. L. (ed.) (1960). *The Biology of Weeds*. Blackwell Science Publishers, Oxford.

Hilbert, D. W., Swift, D. M., Detling, J. K., and Dyer, M. I. (1981). Relative growth rates and the grazing optimization hypothesis. *Oecologia* **51**:14–18.

Hills, T. M. and Peters, D. C. (1971). A method of evaluating insecticide treatment for control of western corn rootworm larvae. *J. Econ. Entomol.* **64**:764–765.

Loring, S. J., Snider, R. J., and Robertson, L. S. (1981). The effects of three tillage practices on Collembola and Acarina populations. *Pedobiol.* **22**:172–184.

Maun, M. A. (1974). Ecological interactions between weed species in agro-ecosystems. *Am. J. Bot.* **61**:46.

McNaughton, S. J. (1979). Grazing as an optimization process: Grass-ungulated relationships in the Serengeti. *Am. Nat.* **113**:691–703.

Moore, J. C., Snider, R. J., and Robertson, L. S. Effects of different management practices on Collembola and Acarina in corn production systems. Part I. Effects of no-tillage and Atrazine. *Pedobiol.*, in press.

Odum, E. P. (1953). *Fundamentals of Ecology*. W. B. Saunders Co., Philadelphia.

Owen, D. F. (1980). How plants may benefit from the animals that eat them. *Oikos* **35**:230–235.

Owen, D. F. and Wiegert, R. G. (1976). Do consumers maximize plant fitness? *Oikos* **27**:488–492.

Perkins, J. H. (1982). *Insects, Experts and the Insecticide Crisis. The Quest for New Pest Management Strategies*. Plenum Press, New York.

Petelle, M. (1982). More mutualisms between consumers and plants. *Oikos* **38**:125–127.

Rabatin, S. C. and Rhodes, L. H. (1982). *Acanlospora biseticulata* inside orbatid mites. *Mycologia* **74**:859–861.

Rice, E. L. and Pancholy, S. K. (1973). Inhibition of nitrification by climax ecosystems. II. Additional evidence and possible role of tannins. *Am. J. Bot.* **60**:691–702.

Sagar, G. R. (1968). Weed biology—a future. *Neth. J. Agric. Sci.* **16**:155–164.

Stinner, B. R., Crossley, Jr., D. A., Odum, E. P., and Todd, R. L. Nutrient budgets and internal cycling of N, P, K, Ca, and Mg in conventional, no-tillage, and old-field systems on the Georgia Piedmont. *Ecology*, in press.

Stinner, R. E., Regniere, J., and Wilson, K. (1982). Differential effects of agroecosystem structure on dynamics of three soybean herbivores. *Ent. Eut.* **11**:538–543.

Thompson, J. N. (1982). *Interaction and Coevolution*. John Wiley & Sons, New York.

Trenbath, B. R. (1981). Plant interactions in mixed crop communities. In *Multiple Cropping*. R. I. Papendick, P. A. Sanchez, and G. B. Triplett (eds.), American Society of Agronomy, Madison, Wisconsin, pp. 129–170.

Tripathi, R. S. (1977). Weed problem—an ecological perspective. *Tropical Ecol.* **18**:138–148.

Decomposition, Organic Matter Turnover, and Nutrient Dynamics in Agroecosystems

David C. Coleman

Natural Resource Ecology Lab and Dept. of Zoology/Entomology
Colorado State University
Fort Collins, Colorado

C. Vern Cole

USDA–ARS
Colorado State University
Fort Collins, Colorado

and Edward T. Elliott

Natural Resource Ecology Lab
Colorado State University
Fort Collins, Colorado

INTRODUCTION

Throughout recorded history, that is, the last 10,000 years, the rise and fall of several civilizations have been linked with changes in the organic matter and nutrient status of agricultural land (Whitney, 1925). Indeed, the development of cropping systems led to the development of early civilizations. Agricultural crops are a principal source of food and fiber upon which our present-day society depends so heavily. Many studies of decomposition processes, that is, loss of dry matter from various residues, relate to problems of organic matter (OM) depletion in soil and the long-term fertility status of various agricultural soils (Martel and Paul, 1974; Bettany et al., 1980; Campbell, 1978).

GLOBAL PERSPECTIVE

The global magnitude of the losses of organic carbon and nitrogen from croplands is not known. It may approach, or be equivalent to, losses of carbon from tropical deforestation or from fossil fuel combustion. Wilson (1978) noted that a major pulse of CO_2-C into the atmosphere, as recorded by delta ^{13}C in bristle-cone pine trees in California, occurred between 1850 and 1890, before the major outputs from coal and petroleum-based industry occurred (Fig. 1). Because of the breakdown of carbon in soil organic matter (SOM), the ^{13}C values for CO_2 in the atmosphere decreased and were recorded in the ligno-cellulose of bristlecone pine (*Pinus aristata*).

The storage of carbon in SOM has been estimated at 30×10^{14} kg (Bohn, 1976) and 14.6×10^{14} kg (Schlesinger, 1977) worldwide. This represents a carbon pool four times greater than either the living biota or the present CO_2 content of the atmosphere (Bolin, 1977; Woodwell et al., 1978). It is clear that inputs to and outputs from the soil OM pool are of considerable significance to the global carbon flows.

With recent concerns about cost-effective management of agroecosystems to minimize fuel and fertilizer inputs, it becomes all the more important to understand fundamental mechanisms of ecosystem function. Is it possible to promote an efficient turnover of principal nutrients, such as nitrogen, phosphorus, and sulfur, in a system that has greater cost constraints than was true only 10 years ago (in North America, and much of the "developed countries")?

One of the possibilities we address is that significant changes in management, such as zero-tillage (Phillips et al., 1980), significantly alter the structural integrity, enhance the long-term stability (*sensu* Holling, 1973), and promote a greater nutrient conservatism consonant with good land management.

Another major point of interest to ecologists is that agroecosystems are useful models to learn about processes of decomposition and nutrient turnover. We will show a number of examples with time scales of from 10–50

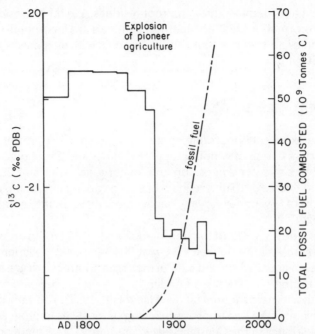

Figure 1.   Delta ^{13}C values of cellulose in bristle-cone pine (*Pinus aristata*) trees of California, showing decrease in values, with influx of soil organic matter-derived CO_2, between 1850 and 1890, due to extensive land clearing and pioneer agriculture (from Wilson, 1978).

years, using field and modeling experiments, which show the exciting possibilities inherent in studying agroecosystem functional processes.

OBJECTIVES

This chapter addresses the following three major objectives.

1. We present aspects of decomposition, OM production and turnover, and OM interaction with nitrogen, phosphorus, and sulfur turnover.
2. We examine management regimes of tillage, fallowing, herbicide application, and so forth, to evaluate effects on the system over a range of 10–50 years.
3. We emphasize the central role of microbial activity in the foregoing processes.

Thus, our overall intention is to present the abiotic and biotic processes that have a marked impact on decomposition of current-year organic matter such as leaf and stem residues and root residues. We then consider subsequent transformations that occur as the materials are incorporated into microbial bodies, further metabolized, and reacted with elements of the soil to become SOM.

HISTORICAL PERSPECTIVE

There is an excellent opportunity to study the prospective (future) and retrospective (past) effects of various cropping regimes and tillage practices upon short-term residue decomposition and longer-term aspects of SOM status. This should enable us to differentiate between varying inputs and outputs of root and shoot OM under varying management regimes (intensive, minimal, or no-tillage).

A major study of soil OM (organic carbon and nitrogen) across nine states and 17 research stations in the U.S. Great Plains showed that approximately 46% of the organic carbon and 42% of the organic nitrogen of grassland soils were lost during 40–50 years of dryland cropping in plots sown to winter wheat, with alternate summer fallow (Haas et al., 1957). The early part of the decay curve and the apparent approaching of a new asymptote are shown for three field sites in the Central Great Plains (Fig. 2). This tendency for loss of organic carbon with cultivation is shown in several examples showing crop rotations or intervening fallow regimes in several countries (Table 1). Inclusion of growing plants in continual rotations (without intervening

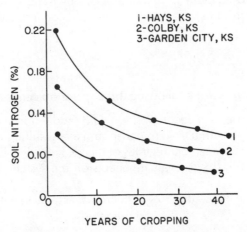

Figure 2. Loss of soil nitrogen concomitant with organic carbon losses in wheat–fallow rotations at three locations in Kansas, over 40 years (from Haas et al., 1957).

Table 1. Management and contents of organic matter in soils.

Soil	Treatment	Organic Matter (%)
Chernozem	Virgin soil	4.33
Lenin State	Old arable	4.00
Silt soil	Grassland 100 yr	7.58[a]
Lincolnshire	Arable 25 yr	2.16[a]
Alfisol	Nontilled	4.52[a]
Nigeria	Tilled	3.38[a]
Red-brown earth	Pasture 30 yr	5.30[a]
Australia	Wheat fallow rotation	2.08[a]
Alfisol	Cover crop	3.14[a]
Nigeria	Weed fallow	2.74[a]

[a] $2 \times \%$ organic carbon (from Tisdall and Oades, 1982).

fallowing) tends to moderate the rate of loss of SOM (Tisdall and Oades, 1982).

There is mounting evidence that certain management practices may significantly alter the rates of losses and even reverse deterioration of many soils. Bauer and Black (1981) have documented some key aspects of these changes (Table 2). They determined the soil properties of 36 fields representing rangeland, wheat production under fallow rotation with recommended stubble mulch practices, and other wheat management practices on a range of soil textures (sandy, medium, fine). The levels of carbon and nitrogen in the surface 46 cm in these soils were significantly affected by the soil texture, reflecting differences in moisture infiltration and storage, the role of clay surfaces in stabilizing organic matter, and other factors (Table 3). Losses

Table 2. Management effects on carbon and nitrogen losses in northern U.S. Great Plains.[a]

Property	Depth (cm)	Haas (1957)	Loss from Virgin Grassland (%) Stubble Mulch	Conventional
Carbon	0–15	41	27	38
	15–30	20	7	14
Nitrogen	0–15	34	23	33
	15–30	16	5	10

[a] From Bauer and Black (1981).

Table 3. A comparison of organic carbon and nitrogen in rangeland, stubble-mulched cropland, and bare fallow cropland, northern Great Plains, U.S.A.

Soil	Rangeland	Recommended Wheat Fallow	Other Wheat Fallow
Organic C[a] (g/m^2)			
Sandy	6684	6017	4392
Medium	9218	6574	6850
Fine	9833	9551	8844
Organic N[a] (g/m^2)			
Sandy	785	757	594
Medium	1039	816	802
Fine	1102	1067	1015

[a] To a depth of 45.7 cm; calculated from data of Bauer and Black (1981).

of both carbon and nitrogen were most pronounced in sandy soils. Losses were not as great on the medium- to fine-textured soils. Differences in management had substantial effects on relative losses in all soils.

MAJOR PROCESSES IN ORGANIC MATTER TURNOVER

First, we will give a definition of decomposition. Campbell (1978) defines decomposition as "breakdown of organic matter to simple organic compounds." Thus, residues added to soils are first broken down to their fundamental organic components by extracellular enzymes produced by heterotrophs. The heterotrophs are in large part bacteria, actinomycetes, and fungi; but there are also contributions from micro-, meso-, and macrofauna. Much of the organic material is ultimately mineralized to CO_2 (Clark and Paul, 1970) by microbial and faunal metabolic activity. Of the soil animals, the layman is most acquainted with the role of earthworms, which are often important in incorporating plant residues into the soil. Other fauna play roles in comminution of materials (Birch and Clark, 1953), digesting microbes, and enhancing decomposition and mineralization. One may envision a successional sequence of heterotrophs, such as the primary microflora which attack the fundamental components of the added carbonaceous substrates; these are succeeded by secondary, tertiary, and so forth, microflora that thrive on the cells and by-products of the primary flora.

Microbial activity is responsible for a very complex array of biochemical processes that break down carbohydrates, lipids, and proteins. Because all organisms follow some relationship to Van't Hoff's rule, there is a change in activity of approximately 2–2.5 with every 10°C rise or fall in temperature (Q_{10}). However, the relationships are complex due to the interplay between the organisms, the decomposing substrate, and the milieu or substrate on which they occur (either on or in the soil), which makes an important difference for the kinetics of decomposition. Several major processes occur in general in all systems, and these will be presented first; then the specifics of the modifying factors will be addressed, particularly as they apply to agricultural ecosystems.

PROCESSES COMMON TO ALL SYSTEMS

Imagine a freshly fallen leaf from a wheat plant that falls to the ground having been fed upon in part by a grasshopper during the growing season. It has a combination of labile, available substances in it and nonlabile ligno-cellulosic constituents, as well. Now envision a similar leaf that remains upright and matures until there is a preponderance of ligno-cellulosic constituents. At harvest time a combine comes through the field, returning the threshed stalks to the field as straw on the ground surface. In both instances there will be a succession of decomposers that break down this material. Given the amount of labile versus nonlabile materials present and similar temperature and moisture conditions, there will be a decomposition rate that reflects the substrate quality (Van Cleve, 1974).

Rates of income and loss of organic matter may be compared mathematically using the negative exponential model (Olson, 1963). For the case of steady litter input, L, the instantaneous rate of change is the limit as Δt and Δx approach zero (Olson, 1963)

$$\frac{dx}{dt} = L - kx$$

The loss rate kx is the product of amount present (x) and the instantaneous fractional loss rate, k.

For the situation in which there is an annual, pulsed input of leaf (shoot) litter, the relationship would be

$$\frac{x}{x_0} = e^{-k}$$

or for any time interval, t

$$\frac{x}{x_0} = e^{-kt}$$

where x_0 = amount of material at time 0;
   x = amount of material at a later time;
   e = the base of natural logarithms;
   k = the fractional loss rate.

For example, in an ecosystem with annual pulses of litter, the accumulation of litter is steadily ascending for the first few years, until a steady-state asymptotic value is reached and litterfall (L) = loss = kx (Fig. 3).

The decomposition rate factor (k) is important because it enables ecologists to compare input and loss functions in a wide array of ecosystems. Values for k have been examined in many ecosystems with low and high production rates (Meentemeyer, 1978; Melillo et al., 1982) and over a wide range of environmental conditions (Fig. 4). Much of the litter decomposition kinetics for agroecosystems would be at high or intermediate levels (Fig. 4). For example, values of k, in various ecosystems, ranged from 0.03 to 5.0 (Table 4). In agroecosystems k is influenced by a variety of factors, which makes it more difficult to interpret than in native ecosystems. For example, the transport of nitrogen in hyphae from the soil to litter varies with tillage practice. Upon tillage, hyphae transporting nitrogen will break but the resi-

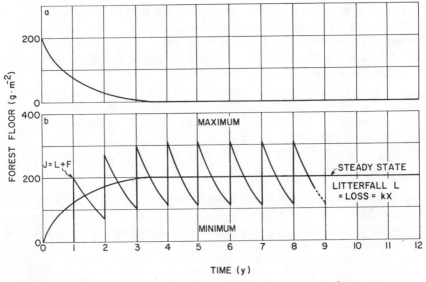

Figure 3. Patterns of inputs and losses of leaf-litter in temperate forest ecosystems. (a) Negative exponential curve for weight loss from a one-time addition; (b) accumulation and losses, with annual pulses (maxima and minima), reaching an asymptote at 200 g m^{-2}. Smooth curve shows conditions of steady income and loss. L = litter input; F = old forest floor at yearly minimum (from Olson, 1963).

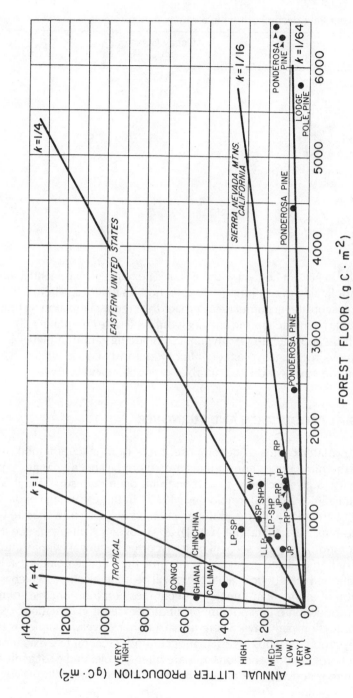

Figure 4. Plots of annual litter production and forest floor standing crops for temperate and tropical coniferous forests. Lines indicate ranges of high (4) to low (1/64) k values per year (from Olson, 1963).

Table 4.   Leaf-litter decomposition in a variety of ecosystems.[a]

	Decay Constant (K value)	
Boreal		
Tundra	0.13	Slow
Taiga	0.13	
Temperate		
Grassland	1.5	
Deciduous forest	0.5	
Tropical		
Equatorial forest	5.0	
Savanna	3.0	Rapid

[a] From Swift et al. (1979).

due will be in greater contact with the soil. The negative exponential model may be applicable in continuous cropped no-till situations, where a steady-state amount of litter accumulates. Perhaps the greatest limitation on the use of k is that it is nonmechanistic and deals only with one component (litter). A more holistic approach to OM dynamics, including various moieties in the soil, employs the negative exponential model to good advantage, as described in the next section on microbial production and turnover.

MICROBIAL PRODUCTION AND TURNOVER

When fresh litter begins to decompose, there is an initial immobilization phase of the nitrogenous constituents until there has been enough respiratory activity to give a change in carbon-to-nitrogen ratio such that it is approximately 20 or 22 : 1 (Bartholomew, 1965). In this early stage of decomposition, particularly if the material such as wheat straw has a wide carbon-to-nitrogen ratio (about 100 : 1), any exogenous nitrogen such as nitrate or ammonium in the soil will be transported by fungal hyphae into the material or immobilized in bacterial cells until the balance point is reached. With continued metabolic activity there will be mineralization, which is the production of inorganic nutrients by decomposition from organic sources. The net production of ammonium or nitrate proceeds via catabolic oxidation or reduction pathways. There are a number of very complex routes, including various gaseous losses, that occur under incomplete oxidation or incomplete reduction conditions and are often accelerated or accentuated where there is considerable organic matter and water present.

The negative exponential model best fits a single homogeneous substrate molecule such as cellulose or chitin, which is being decomposed under relatively defined environmental conditions. Using a more complex resource, total weight loss might be expected to reflect the summation of the decay curves of the individual substrate fractions. This hypothesis was tested by Minderman (1968), who examined the loss rates of principal constituents of forest litter, giving various loss rates ranging from 10% per year for phenols to 99% per year for sugars (Fig. 5). With a slower loss rate, there is an interaction between the components and/or the production of new components, which we now know to be principally microbial in origin. Thus, Clark and Paul (1970) and Paul and Van Veen (1978) noted that by correcting the data on individual component decomposition for resynthesis of secondary metabolites from microorganisms, it approaches the actual summation curve S (Fig. 5).

The negative exponential model is applicable to litter decomposition, but models that include more specific controls are needed to simulate how some of this litter becomes SOM. The SOM model (Parton et al., 1983) represents these processes (Fig. 6). Of the components in this model, two, structural and metabolic, are derived directly from plant litter inputs while the other three components, active soil, slow soil, and passive soil are pools of SOM of increasing recalcitrance, hence longer turnover times (3, 30, and 200 yr, respectively). This model has a relatively low level of resolution and monthly or yearly time steps. It is capable of predicting the buildup or loss of SOM under a variety of residue applications and tillage practices (Parton et al., 1983). A key aspect of the model is the exchange of material between the slow and passive pools.

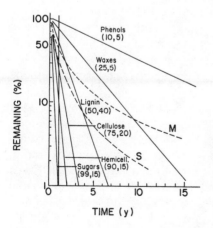

Figure 5. Loss rates of labile substances, such as sugars and hemicelluloses, to very resistant substances (phenols) in soils. S = summation of weighted average of loss rates of materials; M = accumulation and resynthesis of compounds in microbial-derived materials, including humic compounds (from Minderman, 1968).

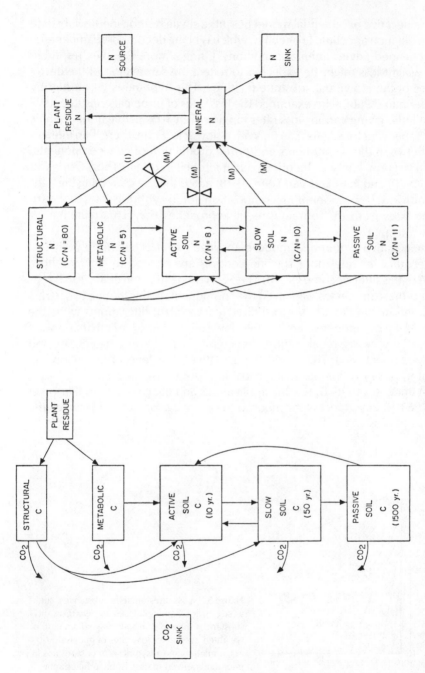

Figure 6. A soil organic matter model showing pools (structural, metabolic, and active) with high turnover rates. Note pathways to mineral N, marked M = mineralization; I = immobilization (from Parton et al., 1983).

Although the SOM model (Parton et al., 1983) does well for simulating events over a period of years, events occurring over a time period of days to weeks, such as microbial growth and death, are not explicitly included in the SOM model. This higher level of resolution is addressed by the PHOE-NIX model (McGill et al., 1981), which simulates carbon and nitrogen flows. Both SOM and PHOENIX models contain flows of carbon and nitrogen. The model can run with daily time steps and includes, among others, the specific processes of root and shoot death, microbial growth based upon Michaelis-Menten kinetics and the internal C : N of the microbes. Fungi and bacteria are treated explicitly with each of these components broken down further to structural (e.g., cell wall) and metabolic components. The model performs well in predicting short-term responses (within a month or year) to perturbations (fertilization, soil fumigation, additional substrate inputs, and cultivation) but also can predict long-term changes in soil nutrient pools.

Stewart and McKercher (1982) present a conceptual diagram (Fig. 7) showing plant root-plant residues feeding into a very complex web of interactions showing, among other pools, stable and labile inorganic and organic forms of phosphorus. Biotic activity plays a central role in the breakdown of plant and other residues. Bacteria and actinomycetes and fungi will be preyed on by a wide array of soil micro-, meso-, and macrofauna. Excreta

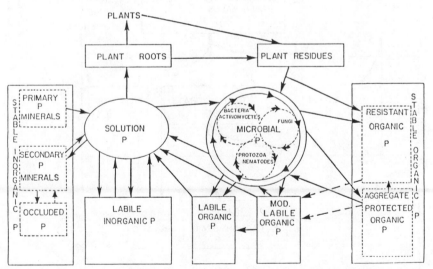

Figure 7. Conceptual model of phosphorus transformations and flows in soil. Stable inorganic and organic forms flow into labile pools, which are moderated (regulated) by microbial–faunal interactions, affecting movement into solution *P*, and hence to plants (from Stewart and McKercher, 1982).

flow to various parts of the system for further cycling or short- or long-term sequestration (Anderson et al., 1981). There are many hierarchical levels on which one can study agroecosystem dynamics, many of which have been formalized into simulation models. Determination of which model is most applicable depends upon the needs of the researcher. It may be advisable that several models at different levels of resolution be applied simulta- neously (Hunt and Parton, in press) for a better understanding of the com- plete process of decomposition and nutrient cycling in agroecosystems.

The complicated biochemical processes that occur in the synthesis of mi- crobial cellular material and resynthesis of various polyphenolic com- pounds, which then lead to "heteropolycondensates" (McGill et al., 1981), are very important. There is an elaborate decomposition sequence ranging from plant detritus to breakdown products (secondary resources) and then humus (Fig. 8). The interaction of primary and secondary substrates and mi- crobial components has been extensively analyzed by McGill et al. (1981). They envisioned a series of components such as leaf litter, soluble compo-

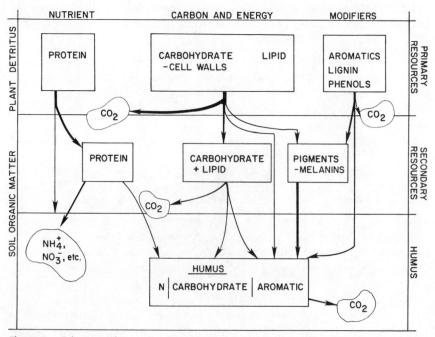

Figure 8. Schematic diagram of decomposition of primary materials leading to secondary resources and finally humic materials (from Swift et al., 1979).

nents, and so-called humads which include humified substances adsorbed onto some of the soil particles.

MICROBIAL-FAUNAL INTERACTIONS

It is helpful to consider the spatial arrangement of the organisms and substrates that they are decomposing. A conceptual diagram (Coleman et al., 1983) shows root, root hairs, root cortex, and the rhizosphere region, and then the actual bulk soil (Fig. 9). Among the biota, there are a sequence of primary decomposers and then fauna feeding upon them in the rhizosphere and bulk soil regions (Fig. 9). As the root grows and sloughs cellular debris (Shamoot et al., 1968), microbes and fauna consume and mineralize these materials and are intensely active in the rhizosphere. Farther back, there is nutrient uptake by the root and associated fungal symbionts (mycorrhizae).

The incredible diversity of organismal components that act upon decomposing substrates has been reviewed by Garrett (1963), Hudson (1968), and Anderson et al. (1981). There is probably no simple sequence of decomposers of very simple labile materials (Swift et al., 1979), but there is a sequence of phases of colonization, exploitation, invasion, and post invasion, which occurs in the decomposition of more resistant and recalcitrant material such as leaf litter, tree branches, twigs, or stumps (Swift et al., 1979; Fig. 10).

One reason for the high variability in the kinds and amounts of organisms colonizing decomposing material is that there is a considerable random or stochastic element to this process that has not been generally appreciated. This random element in microbial and faunal colonization and succession was studied by Fager (1968), who used synthetic logs as models of the decaying wood habitat. Fager used four types of "log": (1) those packed solidly with oak sawdust; (2) those packed solidly with sawdust and lengths of cane to give access as pseudo-boreholes; (3) oak sawdust that had also been mixed with a mixture of bone flour and maize or corn meal (to provide more nutrients than sawdust); and (4) one that had the maize and corn meal and oak sawdust, plus the cane access tubes. The synthetic logs were left out in the forest for periods of 10–16 months. Large and small arthropods were sorted and/or extracted. Fager (1968) found a great diversity of dominant species, but all of the populations had a similar structure with 2 species contributing 50% of the individuals, 3 species contributing an additional 25%, 24 species with 2–74 individuals, and the remaining 12 species represented by single individuals. The population structure was most closely approximated by a mathematical model that assumed that frequency of occurrence of the decomposer fauna is a measure of the probability that a species will

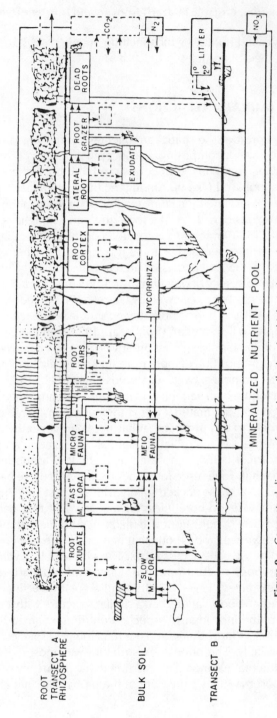

Figure 9. Conceptual diagram of a root in soil, showing biological complexity of microbes, flora, and fauna in the rhizosphere. Transect B depicts bulk soil, with bits of debris and mycorrhizal hyphae, with mesofauna moving into and out of this region (from Coleman et al., 1983).

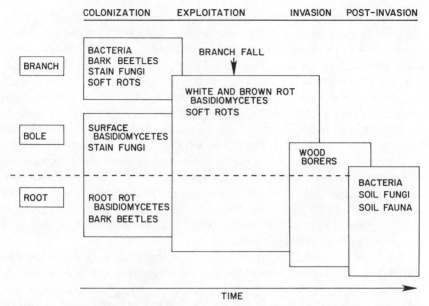

Figure 10. Decomposition sequence of branches, boles, and roots, extending from early phases (colonization) to later ones, such as post invasion. Organisms dominant in these phases are shown (from Swift et al., 1979).

invade a log and that invasion and establishment are in fact random processes. No doubt similar colonization processes occur in agroecosystems.

EFFECTS OF TILLAGE PRACTICES

In contrast to a number of natural systems where there may be a steady rain of leaf litter coming in every year and building up to an equilibrium (Fig. 3), we now consider different aspects of man management (i.e., residue incorporation by plowing versus no plowing) in agricultural ecosystems. There could be some marked changes in decomposition patterns in an agricultural system that has conventional tillage (with deep mold board plowing) or minimal tillage such as stubble mulch or zero tillage in which there is virtually no breaking of the ground surface, but weeds are controlled only with chemical fallow or herbicides. Doran (1980a) showed very different amounts of nitrifiers and denitrifiers associated with leaf litter placed on the surface or buried in the soil of Nebraska corn and wheat fields. These trends showed a 2- to 20-fold increase in nitrifiers and a 3- to 43-fold increase in denitrifiers in zero tillage compared to regular black fallow conventional tillage with plowing. The average ratio of microbial groups between no-tillage and conventional tillage was always greater than 1 in the surface 7.5 cm but less than 1 from 7.5–15 cm except for facultative anaerobes and denitrifiers, which were greater than one at both depths (Doran, 1980b; Table 5). This management regime imposes a wide variety of different physical conditions as well. Thus, the temperature at the litter–soil interface with a thick cover of wheat residue was from 1.5 to 3°C cooler under zero-tillage compared to conventional tillage (Fenster and Peterson, 1979), and the retained moisture was considerably greater (from 2 to 3% by wt) than in the conventional tillage.

Recent studies in agroecosystems in Georgia showed slight increases in diversity of plant root-feeding morphospecies in no-till compared with conventional tillage and increases in densities of earthworms, as well (Stinner and Crossley, 1980). Other studies of microarthropod population changes showed greater numbers of predatory mites and euedaphic Collembola in no-till plots (Moore et al., in press).

We have compared biological and chemical processes under stubble mulch and no-tillage regimes at the Central Plains Research Station, Akron, Colorado. One of the variables of interest to us on the Akron tillage plots is nitrogen since it is limiting in these soils due to the long cropping history. Nitrate accumulated to a greater extent in the fallow than in the cropped rotation (Table 6). Ammonium-N was usually at very low levels (~ 1.0 μg NH_4^+-N $\cdot$ g^{-1} soil), but on one date the concentration reached 8μg NH_4^+-N $\cdot$ g^{-1} soil in the top 2.5 cm of the no-till plots just prior to the highest rate of NO_3^- appearance. More NO_3^- accumulated in the no-till than in the stubble mulch treatments (Table 6). However, it is possible that there was more mineralization in the stubble mulch plots earlier in the year before the first

Table 5. Average ratio of microbial populations between no-tillage and conventional tillage for two soil depths.[a]

	Ratio of Microbial Populations (NT/CT) with Depth	
Microbial Group	0–7.5 cm	7.5–15 cm
Total aerobes	1.35	0.71
Fungi	1.57	0.76
Actinomycetes	1.14	0.98
Aerobic bacteria	1.41	0.68
NH_4^+ oxidizers	1.25	0.55
NO_2^- oxidizers	1.58	0.75
Facultative anaerobes	1.57	1.23
Denitrifiers	7.31	1.77

[a] From Doran (1980b).

Table 6. Distribution of NO_3^--N ($\mu g \cdot g^{-1}$) in the surface layers under the influence of rotation and tillage practice at the Central Great Plains Research Site, Akron, CO, for the summer and fall of 1982.

Rotation	Tillage Practice	Depth (cm)	Sampling Date (1982)				
			8 June	6 July	2 Aug.	23 Aug.	13 Sept.
Cropped	Stubble mulch	0–2.5	4	2	5	8	4
		2.5–5	1	1	5	4	7
		5–10	1	1	5	2	5
		10–20	1	0.5	3	2	3
	No-till	0–2.5	3	3	7	5	5
		2.5–5	2	1	6	3	10
		5–10	1	1	6	2	5
		10–20	1	0.5	3	1	2
Fallow	Stubble mulch	0–2.5	9	12	16	23	10
		2.5–5	7	8	14	21	21
		5–10	5	7	15	9	12
		10–20	8	9	12	10	7
	No-till	0–2.5	9	46	32	84	40
		2.5–5	6	12	12	26	66
		5–10	4	7	13	15	15
		10–20	5	5	10	9	8

sample date, and this mineralized N was moved below our sampling depth (20 cm) as NO_3^--N during a rainfall event. To test this, we sampled to 180 cm to look for deep NO_3^- leaching.

We found more deep NO_3^- in the no-till than in the stubble mulch in the fallow rotation (Table 7). We calculated the total nitrogen accumulation on an areal basis for each treatment using concentration and bulk density information. There was 95 kg $\cdot$ ha^{-1} NO_3^--N under stubble mulch and 167 kg $\cdot$ ha^{-1} under no-till to 180 cm. The no-tillage cultivation practice accumulated 76% more mineral nitrogen than the stubble mulch under the conditions of this experiment during the summer and fall of 1982. Additionally, the nitrogen availability was well timed with plant needs since these plots were planted soon after the NO_3^- had accumulated in the surface. These results agree with those of Powlson and Jenkinson (1981), who found slightly more mineral nitrogen accumulating under direct drilled than in ploughed cultivations.

In conclusion, there seem to be some long-term benefits in terms of mineral nitrogen availability and timing under no-till cultivation practices, possibly as a result of better management of microbial populations. This is in addi-

Table 7.   Depth distribution of NO_3^--N ($\mu g \cdot g^{-1}$ soil) in stubble mulch and no-till management practices at the Central Great Plains Research Site, Akron, CO, on August 23, 1982.

Depth (cm)	Stubble Mulch	No-till
0–2.5	23	84
2.5–5.0	21	26
5–10	9	15
10–20	10	9
20–40	6	9
40–60	7	9
60–90	5	6
90–120	2	4
120–150	1	2
150–180	1	4

tion to minimizing compaction, reducing cost of tillage, maintaining soil organic matter, increasing water use efficiency, curtailing erosion, and protecting soil structure, all of which are often cited as advantages for the use of no-till.

We have seen that the production and turnover of the microbes (the primary decomposers) is central to our understanding of organic matter dynamics and nutrient cycling. The prospects for improved management of microbial populations, hence improved nutrient turnover (possibly optimal timing for mineralization at time of crop growth), will lead to a more cost-effective, efficient agriculture.

REFERENCES

Anderson, R. V., Coleman, D. C., and Cole, C. V. (1981). Effects of saprotrophic grazing on net mineralization. In *Terrestrial Nitrogen Cycles*, Vol. 33. F. E. Clark and T. Rosswall (eds.), Ecol. Bull. (Stockholm), pp. 201–216.

Bartholomew, W. V. (1965). Mineralization and immobilization of plant and animal residues. In *Soil Nitrogen*. W. V. Bartholomew and F. E. Clark (eds.), ASA Monograph 10, American Society of Agronomy, Madison, Wisconsin, pp. 287–306.

Bauer, A., and Black, A. L. (1981). Soil carbon, nitrogen and bulk density comparisons in two cropland tillage systems after 25 years and in virgin grassland. *Soil Sci. Soc. Am. J.* **45:**1166–1170.

Bettany, J. R., Saggar, S., and Stewart, J. W. B. (1980). Comparison of the amounts and forms of sulfur in soil organic matter fractions after 65 years of cultivation. *Soil Sci. Soc. Am. J.* **44:**70–75.

Birch, L. C., and Clark, D. P. (1953). Forest soil as an ecological community with special reference to the fauna. *Q. Rev. Biol.* **28:**13–35.

Bohn, H. L. (1976). Estimate of organic carbon in world soils. *Soil Sci. Soc. Am. J.* **40:**468–469.

Bolin, B. (1977). Changes of land biota and their importance for the carbon cycle. *Science* **196:**613–615.

Campbell, C. W. (1978). Soil organic carbon, nitrogen and fertility. In *Soil Organic Matter.* M. Schnitzer and S. U. Khan (eds.), Elsevier Publishing Company, New York, pp. 173–271.

Clark, F. E., and Paul, E. A. (1970). The microflora of grassland. *Adv. Agron.* **22:**375–435.

Coleman, D. C., Reid, C. P. P., and Cole, C. V. (1983). Biological aspects of nutrient cycling in soil systems. *Adv. Ecol. Res.* **13:**1–55.

Doran, J. W. (1980a). Microbial changes associated with residue management with reduced tillage. *Soil Sci. Soc. Am. J.* **44:**518–524.

Doran, J. W. (1980b). Soil microbial and biochemical changes associated with reduced tillage. *Soil Sci. Soc. Am. J.* **44:**765–771.

Fager, E. W. (1968). The community of invertebrates in decaying oak wood. *J. Anim. Ecol.* **37:**121–142.

Fenster, C. R., and Peterson, G. A. (1979). Effects of no-tillage fallow as compared to conventional tillage in a wheat-fallow system. Research Bulletin 289. Agricultural Experiment Station, University of Nebraska, Lincoln.

Garrett, S. D. (1963). *Soil Fungi and Soil Fertility.* Pergamon Press, Oxford and London.

Haas, H. J., Evans, C. E., and Miles, E. F. (1957). Nitrogen and carbon changes in Great Plains soils as influenced by cropping and soil treatments. Technical Bulletin No. 1164, USDA, Washington, D.C.

Holling, C. S. (1973). Resiliency and stability of ecological systems. *Annu. Rev. Ecol. Syst.* **4:**1–23.

Hudson, H. J. (1968). The ecology of fungi on plant remains above the soil. *New Phytol.* **67:**837–874.

Hunt, H. W., and Parton, W. J. The role of mathematical models in research on microfloral and faunal interactions in natural and agroecosystems. In *Microbial and Faunal Interactions in Natural and Agro-ecosystems.* M. J. Mitchell (ed.), M. Nijhoff, Amsterdam, in press.

Martel, Y. A., and Paul, E. A. (1974). Effects of cultivation on the organic matter of grassland soils as determined by fractionation and radiocarbon dating. *Can. J. Soil Sci.* **54:**419–426.

McGill, W. B., Hunt, H. W., Woodmansee, R. G., and Reuss, J. O. (1981). PHOENIX: A model of the dynamics of carbon and nitrogen in grassland soils. In *Terrestrial Nitrogen Cycles,* Vol. 33. F. E. Clark and T. Rosswall (eds.), Ecol. Bull. (Stockholm), pp. 49–115.

Meentemeyer, V. (1978). Macroclimate and lignin control of litter decomposition rates. *Ecology* **59:**465–472.

Melillo, J. M., Aber, J. D., and Muratore, J. F. (1982). Nitrogen and lignin control of hardwood leaf litter decomposition dynamics. *Ecology* **63:**621–626.

Minderman, G. (1968). Addition, decomposition and accumulation of organic matter in forests. *J. Ecol.* **56:**355–362.

Moore, J. C., Snider, R. J., and Robertson, L. S. Effects of different management practices on Collembola and Acarina in corn production systems. I. The effects of no-tillage and atrazine. *Pedobiologia* (in press).

Olson, J. S. (1963). Energy storage and the balance of producers and decomposers in ecological systems. *Ecology* **44:**322–331.

Parton, W. J., Anderson, D. W., Cole, C. V., and Stewart, J. W. B. (1983). Soil organic matter formation model. In *Nutrient Cycling in Agricultural Ecosystems.* R. Lowrance, R. Todd, L. Asmussen, and R. Leonard (eds.), University of Georgia College of Agriculture Special Publication No. 23.

Paul, E. A., and Van Veen, J. A. (1978). The use of tracers to determine the dynamic nature of organic matter. Transactions, 11th International Congress of Soil Science, Edmonton, Canada, pp. 61–102.

Phillips, R. E., Blevins, R. L., Thomas, G. W., Frye, W. W., and Phillips, S. H. (1980). No-tillage agriculture. *Science* **208:**1108–1113.

Powlson, D. S., and Jenkinson, D. S. (1981). A comparison of the organic matter, biomass, adenosine triphosphate and mineralizable nitrogen contents of ploughed and direct drilled soils. *J. Agric. Sci. Camb.* **97:**713–721.

Schlesinger, W. H. (1977). Carbon balance in terrestrial detritus. *Annu. Rev. Ecol. Syst.* **8:**51–81.

Shamoot, S., McDonald, L., and Bartholomew, W. V. (1968). Rhizodeposition of organic debris in soil. *Soil Sci. Soc. Am. Proc.* **32:**817–820.

Stewart, J. W. B. and McKercher, R. B. (1982). Phosphorus cycle. In *Experimental Microbial Ecology.* R. G. Burns and J. H. Slater (eds.), Blackwells, Oxford. pp. 221–238.

Stinner, B. R., and Crossley, Jr., D. A. (1980). Comparison of mineral element cycling under till and no-till practices: An experimental approach to agroecosystem analysis. In *Soil Biology as Related to Land Use Practices.* D. Dindal (ed.), U.S.E.P.A., Washington, D.C.

Swift, M. J., Heal, O. W., and Anderson, J. M. (1979). *Decomposition in Terrestrial Ecosystems.* University of California Press, Berkeley.

Tisdall, J. M., and Oades, J. M. (1982). Organic matter and waterstable aggregates in soils. *J. Soil Sci.* **33:**141–163.

Van Cleve, K. (1974). Organic matter quality in relation to decomposition. In *Soil Organisms and Decomposition in Tundra.* A. J. Holding, O. W. Heal, S. F. MacLean, Jr., and P. W. Flanagan (eds.), Tundra Biome Steering Committee, Stockholm, Sweden, pp. 311–324.

Whitney, M. (1925). *Soil and Civilization.* D. Van Nostrand, New York.

Wilson, A. T. (1978). Pioneer agriculture explosion and CO_2 levels in the atmosphere. *Nature* **273:**40–41.

Woodwell, G. M., Whittaker, R. H., Reiners, W. A., Likens, G. E., Delwiche, C. C., and Botkin, D. B. (1978). The biota and the world carbon budget. *Science* **199:**141–146.

Agroecosystem Determinants

Robert D. Hart

Winrock International
Morrilton, Arkansas

An analysis of agricultural ecosystems (agroecosystems) requires a combination of the approaches used by systems ecologists, systems engineers, and social scientists. Agroecosystems have all of the properties of natural ecosystems. Energy flow, nutrient cycling, and information processing can be used to explain much of their behavior. However, information processing in agroecosystems and natural ecosystems are very different. For this reason the effect of the environment on agroecosystem and natural ecosystem structure and function is also quite different.

Information processes in natural ecosystems occur at the organism, population, community, and ecosystem level. While many organisms participate in material and energy processes as well as in information processes, some herbivores and most predators, parasites, and pathogens have an impact on the ecosystem that is disproportionate to the amount of energy and materials that they consume (Caswell et al., 1972). These organisms monitor different sources of information, make decisions, and take actions that control material and energy flows. Decisions are made when genetic and behavioral information is compared with information found in the behavior of other organisms or populations, and in vegetation patterns and environmental patterns.

Information processes in agroecosystems occur not only at the organism, population, community, and ecosystem level but at the suprasystem (farm or pastoral system) level as well.

In this chapter I identify five types of factors that affect agroecosystem information processes (the ecological environment, agricultural resources, the household, other agroecosystems, and the state of the agroecosystem). These factors influence two types of decisions (design and control) that, in turn, affect agroecosystem structure and function. Conceptualizing agroecosystem–environment relationships in terms of determinants and decisions leads to specific hypotheses that, when tested, can contribute to the understanding of agroecosystem information processes. The proposed concepts are illustrated using examples from Central America. The practical implications for agricultural research and development and the theoretical implications for agroecosystem and natural ecosystem research are briefly discussed.

CONCEPTS

Like natural ecosystems, agroecosystems are composed of a set of interacting biological and physical components. The presence of specific components and the arrangement of these components in space and time (structure) is determined by the environment. This arrangement of components is capable of processing environmental inputs and producing outputs (function). Environmental factors that affect function, either directly through inputs or indirectly through effects on the structure, are defined as "determinants."

In agroecosystems the farmer takes over many of the roles of the herbivores, predators, parasites, and pathogens. It is noteworthy that these organisms are usually thought of as the natural enemies of farmers. The farmer monitors the environment and the energy and material processes and implements controls that affect the system's structure and function.

Most systems theorists distinguish between a system's environment and its suprasystem, the system in which the phenomenon of interest functions as a subsystem (Miller, 1978). Information processing in agroecosystems cannot be analyzed without considering that agroecosystems are subsystems of farm or pastoral systems. Farms or pastoral systems are composed of a socioeconomic subsystem (the household, resources that are allocated to different activities, etc.) and one or more agroecosystems (Hart, 1981a). Different types of determinants affect different types of decisions that are made within the socioeconomic subsystems of farm and pastoral systems. A first step in analyzing information processes that affect agroecosystem structure and function is to classify determinants and decisions and to formulate and test determinant–decision hypotheses.

Farm or Pastoral Systems

Figure 1 is a diagram of a farm system showing the flows of materials and energy (solid lines) and information (dotted lines) between the ecological and socioeconomic environment and the farm subsystems. Flows of money between the farm and the socioeconomic environment have not been included in the diagram, but it is recognized that, with exception of some subsistence farms, money is used as a medium of exchange for flows of materials and energy between the farm and the socioeconomic environment.

In Figure 1, the information-processing activities required for the management of an agroecosystem are included within the socioeconomic subsystem. Also included in this subsystem is the pool of agricultural resources that can be used as inputs to the different agroecosystems found on the farm and the social and economic components associated with the household.

Decision making within the socioeconomic subsystem of a farm system is an extremely complex process, and its analysis is beyond the scope of this chapter. Agricultural decision making has been studied with mixed success by sociologists, anthropologists, and economists. In general, social scientists have had little success in attempts to use formal decision models to predict economic behavior in nonindustrial countries (Johnson, 1980). However, Gladwin (1976) and others have used simple decision-making flowcharts to explain how farmers consider "aspects" such as profit, risk, capital, and knowledge to decide which crops or cropping systems to plant. In general, social scientists studying agricultural decision making have focused on farmers' objectives. There has been little research on how these socioeconomic objectives are combined with what is ecologically feasible. How farmers combine crops or livestock in designing an agroecosystem and how they make specific management decisions have been studied even less.

For purposes of this analysis farm system decision making has been conceptualized as three subprocesses: (1) definition of socioeconomic objectives, (2) design of a farm management plan that includes agroecosystem plans for each agroecosystem, and (3) control decisions that are made after the design has been implemented and the system is functioning. As suggested earlier, the first process (definition of objectives) is essentially a suprasystem decision and beyond the scope of this chapter. In the farm system diagram (Fig. 1) socioeconomic objectives are assumed to be an output from the household. The second and third processes (design and control) are part of the information-processing subsystem of an agroecosystem and must be considered in any analysis of how environmental factors affect agroecosystem structure and function.

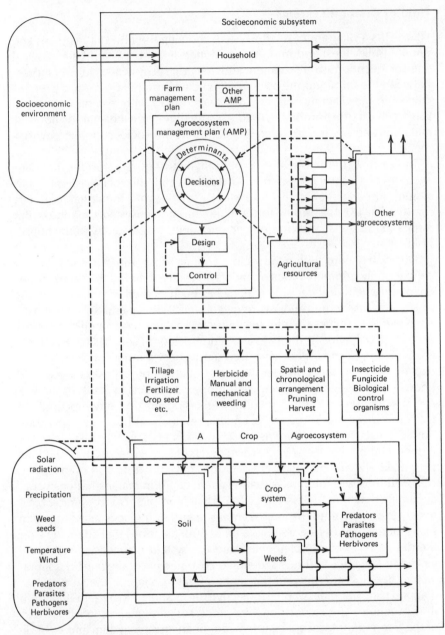

Figure 1. The flow of energy and materials (solid lines) and information (dotted lines) in a farm system. The management of an agroecosystem is conceptualized as a series of decisions based on different types of determinants.

Design and Control Decisions

Agroecosystem design and control decisions are usually cyclical. Design decisions are made during planning, while control decisions are made during implementation. Identification of these cycles is relatively easy in annual-crop agroecosystems, more difficult in livestock-based agroecosystems, and most difficult in perennial-crop agroecosystems.

Design decisions are usually seasonal. For example, during the winter in temperate climates or the dry season in the wet-dry tropics, a farmer or pastoralist decides which components to include in the agroecosystem, how they will be arranged in space and time, and the quality, quantity, and timing of inputs and outputs. The product of this decision-making process is a model (usually mental) of the structure of the agroecosystem and a predicted function. Depending on the level of confidence in this prediction, the pastoralist or farmer may include in the design the possibility of implementing optional control activities.

Control decisions are continuous. Once the structure has been designed and the system starts to function (processing inputs and producing outputs), the control phase begins. During this phase the predicted, real, and projected states of the agroecosystem are continuously evaluated. Some control decisions are programmed and are triggered by the calendar (based on past experience). Other control decisions may require that a combination of factors meet a combination of criteria. For example, a farmer may design a maize agroecosystem that includes a control decision to plant maize only when the calendar, rainfall, and labor availability meet the criteria of "the first week in May, 25 mm of rainfall during the last week in April, and 5 mandays of available labor."

Some control decisions are not programmed during the original design. When the real behavior of an agroecosystem is very different from the expected, unplanned control decisions must be made. This may require an unplanned return to the design phase (starting over), or in some cases information processing may be "returned" to the herbivores, predators, parasites, and pathogens (giving up).

Types of Determinants

Agroecosystem design and control decisions are made on the basis of one or more of the following types of determinants:

1. The Ecological Environment. Some environmental factors are relatively static (soil physical properties, soil fertility, solar radiation); others are very dynamic and less predictable (rainfall patterns, temperature in temperate climates, invasion of pests). These factors affect the agroecosystem (a) directly through the availability of energy and material inputs and (b) indirectly, when design and control activities are triggered by environmental factors.

2. Agricultural Resources. These factors include land, labor, capital, management capability, and agricultural inputs, such as agrochemicals and animal and machine energy. Factors that affect resource availability, and therefore indirectly affect the agroecosystem, are (a) the socioeconomic environment, (b) the household, and (c) the state of other agroecosystems. The household and other agroecosystems can also directly determine design and control decisions and for this reason are listed below as "determinants."

3.   The Household. The principal direct effect (as opposed to indirect effects through other agroecosystems or the resource pool) of the household on the agroecosystem is through design decisions. The definition of socioeconomic objectives (e.g., minimize risk) or the demand for specific outputs (e.g., goat milk) can greatly affect the design process. Control decisions can also be affected by sudden changes in household demand; for example, animals may be sold earlier than planned because of an unexpected economic crisis.

4. Other Agroecosystems. The principal direct effect (as opposed to indirect effect through the resource pool) of other agroecosystems is through the flow of materials or energy among agroecosystems. Examples include the use of crop residue from one agroecosystem to feed livestock in another and the use of animal energy or manure from one agroecosystem to produce crops in another.

5. The Agroecosystem. As suggested earlier, control decisions are triggered by comparing the real and predicted state of the agroecosystem. The cumulative experience with the performance of the agroecosystem is also an important information source for design decisions.

Determinant-Decision Hypotheses

Specific determinants can be combined with decisions to form determinant–decision hypotheses. In the analysis of an agroecosystem all of the control or design decisions that alter a material and energy process must be identified, and specific hypotheses as to which factor or combination of factors determine the decisions must be identified and tested.

Specific determinant–design or determinant–control hypotheses can be tested separately, or a complete agroecosystem management plan, with all the design and control decisions and determinants that trigger them, can be tested. As in the validation of any model, the same data used to identify the hypothesis cannot be used to test it. Validation can be done by beginning with decisions and predicting the existence of specific levels of determinants or by beginning with the determinants and predicting the existence of specific agroecosystems.

It is also possible to identify determinant–control and determinant–design hypotheses by deductive logic or intuition. In a heterogeneous geographic area where the hypothesized determinants exhibit high enough variation so that decisions, and therefore the structure and function of agroecosystems, can be expected to vary, both determinant and decision data can be collected and analyzed to see if the predicted determinant–decision relationships are statistically significant.

AGROECOSYSTEM DETERMINANTS IN CENTRAL AMERICA

The Centro Agronomico Tropical de Investigacion y Ensenanza (CATIE) has conducted agricultural research in a broad range of environments in Central America and is specifically interested in the relationship between environmental factors and farmers' choice of cropping system and the management of specific cropping systems (Moreno, 1980). The primary objective of most of CATIE's research is to identify potential alternatives to what farmers are presently doing. This objective has required an approach that includes an analysis of predominant agroecosystems (Hart, 1981a), the farm systems in which they function (Hart, 1981b), and the factors that determine design and control decisions. Examples from various CATIE studies are cited below.

Example of Design Determinants

Diaz (1982) studied the relationship between various factors and intercropped maize and beans (M + B); maize and sorghum (M + S); and maize, beans, and sorghum (M + B + S). The study was based on a survey of 368 farmers in Honduras. Data were collected on cropping systems, farm systems, and soils and climate. Hypotheses relating potential determinants to design decisions were evaluated by analyzing the percentage of farmers using a specific cropping system (e.g., M + B) at different intervals along environmental gradients and by using principal component and multiple regression techniques to determine the relationship between potential determinants and different management practices (e.g., soil fertility and spatial arrangement of crops).

The study cited above has not been completely analyzed, but some examples from a preliminary analysis illustrate how different types of determinants affect different design decisions. An ecological determinant that was directly associated with the presence of either M + B, M + S, or M + B + S was altitude above sea level (a proxy for temperature). The M + S cropping system was found between sea level and 750 m above sea level, with the highest concentration between 0 and 250 m. The M + S + B system was found between 0 and 1250 m but was concentrated between 500 and 1000 m. The M + B system was found over a wide range in altitudes (0–2000 m); however, different species of beans were intercropped with maize at different altitudes. *Vigna spp.* was planted below 500 m, *Phaseolus vulgaris* was planted at intermediate altitudes, and *P. coccineous* was planted above 1750 m.

A resource determinant (labor availability) affected chronological arrangement. The requirements of other agroecosystems were identified as determinants of the quality and quantity of biomass removal. Of the farmers with M + S or M + B + S, 80% fed maize and sorghum stover to cattle, while only 40% of the farmers with M + B had cattle and only 53% used the maize stover for feed.

Examples of Control Determinants

In spite of the fact that agricultural research has been done with many different agroecosystems in Central America, there have been very few studies of how farmers make control decisions. It is often assumed that farmers decide what to do and then simply implement the package they have designed. In reality, the product of the design process is usually a complex set of options that are triggered by different determinants. A few examples from Central America are cited below.

A group of extension agents in El Salvador were asked how farmers make decisions in the management of intercropped maize and sorghum (Hart, 1980a). Many decisions are triggered by a combination of factors. For example, planting is done either in May after it has rained twice within a two-week period or in June, after one rainfall. Weeding, fertilizer application, hilling-up maize, and planting sorghum are "designed" to occur sequentially beginning three weeks after planting maize, but in reality are done within a two- to five-week period depending on rainfall. If rainfall occurs during the flowering of maize, maize grain yields will probably be adequate, and the intercropped sorghum may be removed before it is mature and fed to livestock, instead of letting it produce grain for human consumption.

An example of farmers implementing control decisions that completely alter the structure and function of an agroecosystem can be cited from Honduras (Hart, 1981b). Farmers in the community of Yojoa usually plant double-cropped maize (maize followed by maize) and a rice–bean rotation (rice followed by beans). The rice–bean rotation is "designed" to be planted in

May; however, if the rainy season is delayed until June, maize is planted instead. A control decision (what crop to plant) made on the basis of rainfall (an ecological determinant) produces a major short-term change in the agroecosystem. This decision can, in turn, affect the farm system (agricultural resources, other agroecosystems, etc.) and the long-term design decisions that are made during the next dry season.

PRACTICAL AND THEORETICAL IMPLICATIONS

An understanding of how the environment and suprasystem affect agroecosystem structure and function, through the interaction of information and material and energy processes, is of both practical and theoretical value. Agricultural scientists doing applied research that interfaces directly with development activities are interested in how the farmers' systems operate. Agricultural scientists doing theoretical research are interested in identifying general principles that can make applied research more efficient.

Ecologists have often conceptualized agroecosystems as natural ecosystems that have been "subverted" or "degraded" by man, overlooking the fact that, as ecosystems with information processing concentrated in one species, they may be ideal for the study of ecosystem information processes. An understanding of the theoretical basis for information processes in agroecosystems will contribute to the theoretical understanding of natural ecosystems as well as agroecosystems. The practical and theoretical implication of conceptualizing agroecosystems in terms of decision determinants are discussed below.

Farm and Pastoral System Research

Agricultural research institutions in some third-world countries have recently shifted emphasis away from commodity-oriented research towards research directed at production systems. The change is based on a recognition that agricultural technology must be designed to fit existing farming systems or pastoral systems. Two important implications are that (1) the suprasystem must be analyzed since the evaluation of any agroecosystem changes must be done on the basis of the performance of the farm or pastoral system and (2) the direct participation of the farmer or pastoralist can no longer be just an abstract goal; it is now a necessity.

Agroecosystem research must be an integral part of any farm or pastoral system research program. An understanding of how determinants affect design and control decisions is needed to understand how a specific farmer's or pastoralist's present system works, so that alternatives can be designed and tested. The "alternatives" are usually agroecosystem-level recommendations rather than specific inputs, varieties, and so forth. The most logical way to organize the information that is the product of applied agroecosystem research is as an agroecosystem management plan. This plan, by definition, must include design and control decisions and the determinants that trigger these decisions.

One criticism of farm and pastoral system research is that it tends to be site-specific. It is possible to study one specific type of farm system in one small geographic area and successfully identify potential alternatives, but it is not economically feasible to do that kind of research with every community of farmers or pastoralists. The question is, how can agricultural technology be transferred between similar environments so that research done in one area can be applied in another? When this general question is addressed, the first specific question is: Which environmental factors should be considered in identifying "similar" environments? An obvious answer is: The determinants of agroecosystem design and control decisions. If a factor is not a determinant, then two areas that differ considerably as to that factor can still be "similar enough" for an agroecosystem management plan to be transferred from one area to another. If a factor is a determinant, the levels of the factor in two areas must be sufficiently similar so that farmers in both areas make the same decision when confronted with that factor.

Agricultural Systems Modeling

Van Dyne and Abramsky (1975) reviewed and evaluated the literature on agricultural system models. Most agricultural models are either simulation models (differential, difference, algebraic, or matrix equations) or optimization models (linear, nonlinear, or dynamic programming). Simulation and optimization models are seldom combined, and there is a distinct tendency to separate decision-making processes from biological processes. Natural ecosystem models can depict most flows between components as a constant coefficient times a state variable; agroecosystem models must include the possibility of not only sudden changes in coefficients that will affect the function of the system, but also of sudden changes in the structure of the system as entire populations are removed (e.g., when crops or livestock are harvested).

Clymer (1972), in a discussion of the next generation of models in ecology, states that man is almost always included in agricultural ecosystem models as "an exogenous entity" or an "off-stage forcing function" rather than as a component and suggests that man should be included "in every ecosystem that is modeled as an applied problem." If this suggestion were followed, the almost complete lack of tested determinant hypotheses would be immediately obvious.

At a minimum it would seem that agroecosystem models should include control decisions. This can be done quite easily by combining differential equations to simulate material and energy processes and logic circuits to simulate information processes, as in the approach developed by H. T. Odum (1972). Using this approach, a coefficient can be changed when a forcing function or state variable (or combinations) satisfy certain criteria. The problem, once again, is a lack of tested hypotheses as to which determinants affect specific decisions. At what level does an environmental factor, an agricultural resource, a household demand, the state of another agroecosystem, or a state variable trigger a decision? For a model to even begin to simulate reality, the logic circuit must include the possibility of implementing controls that are triggered by any (or combinations) of these five types of determinants.

In a model of a bean, maize, and cassava agroecosystem (Hart, 1974), the decision of when to weed was assumed to be determined by "time after planting." However, in the validation process using data from an environment with higher weed growth, it immediately became obvious that the amount of weeds (state of the agroecosystem) must also determine when to weed, or the weeds would out-compete the crops. In reality, farmers decide when to weed, or to weed at all, on the basis of time after planting, amount of weeds in the field, and soil moisture availability. A hypothesis as to how these factors determine the control decision must be formulated and tested before it can be included in an agroecosystem model.

Natural Ecosystem Information Processes

In the preceding discussion the differences between information processes in agroecosystems and natural ecosystems have been emphasized. The fact that many of the information-processing roles of herbivores, predators, parasites, and pathogens in natural ecosystems are appropriated by man in agroecosystems does not make the principles and theoretical concepts derived from the study of one type of ecosystem of no value in the study of the other. In fact, the study of agroecosystems has been based almost completely on principles and concepts derived from the study of material and energy processes in natural ecosystems. It is possible that what is learned from the study of information processes in agroecosystems may be of value in the study of natural ecosystems.

The study of agroecosystems could contribute to an understanding of many of the issues of interest to theoretical ecologists. To illustrate the potential contribution of agroecosystem studies to population, community, and ecosystem ecology, a few comments related to the areas of population inter-action, natural succession, and ecosystem optimization are presented be-low.

Herbivores, predators, parasites, and pathogens are part of nutrient cy-cles and energy flows, but they also use genetic and behavioral information to monitor environmental and community patterns and take actions that af-fect the long-term evolution (design) and short-term function (control) of the ecosystem. For example, a change in the environment can trigger bird mi-gration, and a change in the diversity of vegetation can trigger insect popula-tion outbreaks. The determinant–decision concept proposed for the study of agroecosystems may be applicable in the analysis of how genetic, behav-ioral, vegetation pattern, and environmental pattern information is pro-cessed by herbivores, predators, parasites, and pathogens.

Cybernetics and information theory is often used to describe ecosystems structure and function (Margalef, 1968). However, Johnson (1970) suggests that "it is doubtful whether information theory has offered experimental bi-ologists anything more than vague insights and beguiling terminology" and concludes that what is needed is a qualitative factor that considers biological relevance. In discussions of ecosystem cybernetics, predator–prey (plant–animal and animal–animal) relationships are often presented as examples of negative and positive feedback control. However, predator–prey relation-ships are usually modeled using variables such as number of individuals with rates determined by population density, self-inhibiting effects, and re-source availability (e.g., Smith, 1974); information processes are not em-phasized.

In agroecosystems there is an interesting relationship between the amount of information that goes into seed dispersal, the information content of the vegetation pattern that is produced, and the amount of information required by herbivores to maintain a specific herbivory rate. A farmer plants crops in rows (high order, low information) so harvesting will be easier (re-quire less information). Unfortunately for farmers, natural herbivores also require less information, and natural herbivory is also high. Perhaps theories relating information to energy could be tested by measuring the energy ex-pended by herbivores as the information content of vegetation patterns are changed.

Interest in using biological controls in agroecosystems and in the study of their effect on community stability (see discussion by Watt, 1965) would seem to also offer an opportunity to study predator–prey relationships in terms of information processes. For example, it may be possible to quantify the effect of herbivores, predators, parasites, and pathogens on plants in in-formation terms by measuring the difference between the amount of infor-mation required by farmers using minimum biological control (high chemical inputs), introduced biological control (imported organisms), or natural biological control (managing existing herbivores, predators, parasites, and pathogens).

An understanding of natural succession may provide useful information for the design of agroecosystems (Hart, 1980b), but it may also be true that the study of what determines a farmer's decision to change cropping systems may contribute to an understanding of natural succession. E. P. Odum (1969) describes natural succession as an orderly process controlled by the community in which the physical environment determines the pattern and the rate of change and sets the limits as to how far development can go. The process culminates in a stabilized ecosystem with high information content. Its "strategy" (objective) is to maximize internal control of homeostasis.

Questions that are asked in both the study of agroecosystems and successional ecosystems are: How does the community control the process? How does the environment determine patterns, change, and limits? What is the general "objective"? The study of design and control decisions and their determinants in slash-and-burn (successional) agroecosystems might yield some interesting insights into information processes during natural succession. In these agroecosystems farmers are continuously changing community structure. Crop species and cropping patterns that are planted after cutting and burning the forest are changed as soil fertility decreases and weeds, herbivores, predators, parasites, and pathogens increase.

Spedding (1975) defines an agricultural system as an ecosystem with a purpose. It is interesting to note that to explain natural ecosystem behavior, ecologists have also used teleological concepts such as "maximizing gross production" (E. P. Odum, 1971) and "maximizing power" (H. T. Odum, 1972). Evolution is said to operate "as a mechanism that maximizes fitness, or the relative contribution of a genotype to future generations" (Cody, 1974). The study of the relationship between a farmer's or pastoralist's objectives (maximizing output per unit of resource input, minimizing risk, etc.) and the type of agroecosystem that is designed may even provide insights into the "objective" of natural ecosystems.

SUMMARY

An important difference between agroecosystems and natural ecosystems is that many of the information processes that are controlled by herbivores, predators, parasites, and pathogens in natural ecosystems are controlled by man in agroecosystems. The decisions that directly affect the structure and function of an agroecosystem can be subdivided into design decisions (planning) and control decisions (implementation). These decisions are affected by the state of the agroecosystem and determinants from the ecological environment and the suprasystem (farm or pastoral system). Included within the suprasystem determinants are other agroecosystems, resource availability, and the household. To analyze an agroecosystem, hypotheses relating specific determinants to specific decisions must be identified and evaluated. Examples of determinant–decision hypotheses that have been tested in Central America include temperature to cropping system selection (a design decision) and rainy season onset to selection of the first crop in a rotation system (a control decision). An understanding of information processes in agroecosystems is of practical value for agricultural research and development and of theoretical value in identifying principles that can contribute to a general understanding of the interaction between energy, material, and information processes in natural ecosystems as well as in agroecosystems.

REFERENCES

Caswell, H., Koenig, H. E., and Resh, J. A. (1972). An introduction to systems science for ecologists. In *Systems Analysis and Simulation in Ecology*, Vol. 2. B. C. Patten (ed.), Academic Press, New York, pp. 1–78.

Clymer, A. B. (1972). Next-generation models in ecology. In *Systems Analysis and Simulation in Ecology*, Vol. 2. B. C. Patten (ed.), Academic Press, New York, pp. 533–569.

Cody, M. L. (1974). Optimization in ecology. *Science* **183:**1156–1164.

Diaz, R. E. (1982). Caracterizacion y relaciones ambiente-manejo en sistemas de frijol y sorgo asociados con maiz en Honduras. M.S. thesis. Universidad de Costa Rica, Centro Agronomico Tropical de Investigacion y Ensenanza. Turrialba, Costa Rica.

Gladwin, C. A. (1976). A view of the plan puebla: An application of hierarchical decision models. *Am. J. Agric. Econ.* **58:**881–887.

Hart, R. D. (1974). The design and evaluation of a bean, corn, and manioc polyculture cropping system for the humid tropics. Ph.D. dissertation. University of Florida. Gainesville, Florida. University Microfilms: 75-19, 431. Ann Arbor, Michigan.

Hart, R. D. (1980a). *Agroecosistemas: Conceptos Basicos*. Centro Agronomico Tropical de Investigacion y Ensenanza. Turrialba, Costa Rica.

Hart, R. D. (1980b). A natural ecosystem analog approach to the design of a successional crop system for tropical forest environments. *Biotropica*, Suppl. Tropical Succession **12:**73–82.

Hart, R. D. (1981a). An ecological systems conceptual framework for agricultural research and development. In *Readings in Farming Systems Research and Development*. W. W. Shaner, P. F. Philipp, and W. R. Schmehl (eds.), Westview Press, Boulder, Colorado, pp. 44–58.

Hart, R. D. (1981b). One farm system in Honduras: A case study in farm systems research. In *Readings in Farming Systems Research and Development*. W. W. Shaner, P. F. Philipp, and W. R. Schmehl (eds.), Westview Press, Boulder, Colorado, pp. 59–73.

Johnson, A. (1980). The limits of formalism in agricultural decision research. In *Agricultural Decision Making*. P. F. Barlett (ed.), Academic Press, New York, pp. 19–43.

Johnson, H. A. (1970). Information theory in biology after 18 years. *Science* **168:**1545–1570.

Margalef, R. (1968). *Perspectives in Ecological Theory*. University of Chicago Press, Chicago.

Miller, J. G. (1978). *Living Systems*. McGraw-Hill, New York.

Moreno, R. A., ed. (1980). *Localizacion de Sistemas de Producion de Cultivos en Centroamerica*. Centro Agronomico Tropical de Investigacion y Ensenanza. Turrialba, Costa Rica.

Odum, E. P. (1969). The strategy of ecosystem development. *Science* **164:**262–270.

Odum, E. P. (1971). *Fundamentals of Ecology*. W. B. Saunders, Philadelphia.

Odum, H. T. (1972). *Environment, Power, and Society*. Wiley-Interscience, New York.

Smith, J. M. (1974). *Models in Ecology*. Cambridge University Press, Cambridge, Massachusetts.

Spedding, C. R. W. (1975). *The Biology of Agricultural Systems*. Academic Press, London.

Van Dyne, G. M. and Abramsky, Z. (1975). Agricultural systems models and modelling: An overview. In *Study of Agricultural Systems*. G. E. Dalton (ed.), Applied Science Publications, London, pp. 23–106.

Watt, K. E. F. (1965). Community stability and the strategy of biological control. *Can. Ent.* **97:**887–895.

Energy Flow in Agroecosystems

David Pimentel

Department of Entomology and
Section of Ecology and Systematics
Cornell University
Ithaca, New York

INTRODUCTION

Before the development of agriculture and the manipulation of natural ecosystems to produce certain food plants and animals, wild plants and animals were the only source of food for humans. Originally, humans, like other animals, were an integral part of the natural ecosystem and depended upon the energy flow in these systems for their survival. For most of the time since they evolved, humans have "lived off the land" as hunter–gatherers. Because in primitive times the density of humans to land was low, they were able to remove a small portion of the energy produced by the ecosystem for their food without substantially reducing the productivity of the ecosystem. Under these conditions the ecological system was adequate when the human population in the world numbered less than 10 million (Coale, 1974).

Slowly, as the human population increased to or above the carrying capacity of natural ecosystems, a more structured or planned method of ensuring an adequate food supply, or agriculture, evolved. At first agriculture was simple and probably entailed sowing seeds, left over from their food collecting, on slightly disturbed soil. The eventual harvest was that which survived weed competition, insect pests, plant pathogens, and bird and mammal attacks. Although such a harvest yield was small by present standards, the quantity of desirable food was greater than that obtained by a similar amount of energy invested in searching wide areas of natural ecosystems for food. Thus, a slight alteration of the ecosystem and encouragement of desirable vegetation types provided more food for the same effort that had been expended in hunting and gathering.

Gradually, more intense alteration of ecosystems, employing slash-and-burn agricultural technology, developed and increased food yields. With greater effort and investment in human energy, seedbeds were improved and losses to pests, especially weeds, were reduced. The greater investment of human energy to manage the ecosystem was rewarded with larger food-crop yields and energy value in the form of food than previously possible.

Basically, all agriculture is manipulation of an ecosystem to produce plants and animals needed or desired by humans for food and fiber. Encouraging certain species of plants and animals, while concurrently employing methods to discourage the growth of unwanted species, requires significant inputs of human, animal, wood, and fossil energy. Thus, it is not surprising that as alteration of the ecosystem increased and control of the environment intensified, energy inputs, especially fossil energy, also increased.

Some of the major changes that have taken place in human and fossil energy inputs as agriculture has evolved and intensified are reviewed here. Also examined is the harvest of solar energy by various agricultural systems, including human-powered, animal-powered, and tractor-powered systems. Finally, alternatives that reduce fossil energy inputs while maintaining the high efficiency of converting solar energy into biomass will be assessed.

SOLAR ENERGY

The foundation of the total life system rests on the unique capacity of plants to convert solar energy into stored chemical energy. This captured energy is then utilized by the consumers. The success of agriculture is measured by the amount of biomass energy fixed as a result of manipulating plants, land, water, and energy (both solar and fossil) resources.

The solar energy reaching 1 ha during the year averages about 14×10^9 kcal (Reifsnyder and Lull, 1965). During an average four-month summer growing season in the temperate region nearly 7×10^9 kcal reach the agricultural hectare. Under favorable conditions of moisture and soil nutrients, corn is considered one of the most productive food and feed crops per unit area of land (Pimentel, 1980). For example, high-yielding corn grown on the good soils of Iowa can produce about 7000 kg/ha of corn grain plus another

7000 kg/ha of biomass as stover. Converted to heat energy this totals 69 × 10^6 kcal and represents about 0.5% of the solar energy reaching the hectare during the year (1% during the growing season).

For other crops the efficiency conversion is much less than for corn. For example, potatoes with a yield of 40,000 kg/ha have a dry weight of about 8000 kg/ha. Based on total biomass produced of 12,000 kg/ha and an energy value of 50 × 10^6 kcal, potatoes have a 0.4% efficiency of conversion. Or a wheat crop, yielding 2700 kg/ha of grain, produces a total of 6750 kg biomass/ha which has an energy value of 28 × 10^6 kcal. The conversion efficiency of sunlight into biomass in this system is only 0.2%.

Although all these efficiencies for conversion are low relative to the total amount of solar energy reaching a hectare of land, they are still 2–5 times greater than the average conversion efficiency of natural vegetation, which in the United States is estimated to be about 0.1% efficient in solar energy conversion to biomass (Pimentel et al., 1978).

ENERGY INPUTS FOR CORN PRODUCTION

Continuing the analysis of corn production, several other energy inputs are necessary to manipulate an ecosystem and produce corn biomass instead of natural vegetation biomass. In this analysis three corn production systems will be compared: hand-powered, tractor-powered, and horse-powered.

Corn produced by hand in Mexico employing swidden or cut–burn agricultural technology requires only one man with an axe, hoe, and some corn seed (Table 1). The energy input for the human power is calculated at 3500

Table 1. Energy inputs in corn (maize) production in Mexico using only manpower.

Item	Quantity/ha	kcal/ha
Labor	1,144 hr[a]	500,500
Axe and hoe	16,570 kcal[b]	16,570
Seeds	10.4 kg[b]	36,608
Total		553,678
	output	
Total yield	1,944 kg[a]	6,901,200
kcal output/kcal input		12.5

[a] Lewis, 1951.
[b] Estimated.

kcal per man day and is assumed to come directly from eating the corn grain produced. Since a total of 1144 hours of labor input goes into the system, a total of 143 man days of fuel is required, or 500,500 kcal. This is the single largest energy input for this production system. The labor fuel input would be increased if other inputs needed to sustain the worker were added such as clothing, housing, transport, schooling, police, and fire protection. Then the energy requirements needed by the worker's family could also be included as a necessary input.

Under certain circumstances for swidden agricultural production, one might also include the energy produced by burning the trees and other vegetation that was present on the site before burning. However, for the present the system will be kept relatively simple and include only the kilocalories removed from the corn grain required to fuel the manpower input.

When the energy for making the axe and hoe and producing the seed is added to human power, the total energy input needed to produce corn by hand is only about 553,678 kcal/ha (Table 1). With a corn yield per hectare of 1944 kg or 6.9 million kcal, the output–input ratio is about 12.5 : 1 (Table 1).

In this system the fossil energy input for production is only for the axe and hoe. Based only on the fossil energy input of 16,570 kcal, the output–input ratio is about 422 kcal of corn produced for each kilocalorie of fossil fuel expended.

The energy flow in tractor-powered agriculture is distinctly different from that of man-powered agriculture. Corn production in the United States is typical of the heavy reliance on machinery for power. The total manpower input is dramatically reduced compared to the hand-powered system in Mexico and averages only 12 hours (Table 2) or about 1/100th that of the hand-powered system. The input for labor in the U.S. system is only 7000 kcal of food energy for the growing season or substantially less than that of any other input in production.

Balanced against this low manpower input is the significant increase in fossil energy input needed to run the machines that replace man. In 1980 the fossil fuel energy inputs required to produce a hectare of corn averaged about 8.4×10^6 kcal/ha or the equivalent of about 840 liters of oil (Table 2). Then based on a corn yield of about 7000 kg/ha, or the equivalent of 24.5×10^6 kcal energy, the output/input ratio is 2.9 : 1. Note, the fossil energy input in the system represents about 12% of the solar energy captured by the corn crop (69×10^6 kcal).

The energy flow in corn production can be examined another way, that is, substituting horse power for fossil fuel and the tractor. To do this, the tractor power in Table 2 was removed and 120 hr of horse power plus 120 hr of

Table 2.   Energy inputs per hectare for corn production in the United States.[a]

Item	Quantity/ha	kcal/ha
Labor	12 hr	7,000
Machinery	55 kg	990,000
Gasoline	16 l	264,000
Diesel	77 l	881,500
LP gas	80 l	616,400
Electricity	33.4 kwh	95,500
Nitrogen	151 kg	2,220,000
Phosphorus	72 kg	216,000
Potassium	84 kg	134,000
Lime	426 kg	134,400
Seeds	18 kg	445,500
Insecticides	1.4 kg	119,950
Herbicides	7 kg	777,500
Drying	7,000 kg	1,437,800
Transportation	200 kg	51,200
Total		8,390,750
	Output	
Total yield	7,000 kg	24,500,000
kcal output/kcal input		2.9

[a] Pimentel and Burgess, 1980.

needed manpower were included. An estimated 136 kg of corn and 136 kg of hay would be required to feed a 682 kg (1500 lb) horse (Morrison, 1956). The corn grain would come directly out of the corn produced. The hay required to support the horse power for 1 ha of corn would have to be produced on hayland, but this would represent only 0.2 ha. Thus, based on the current high yields of corn and hay per hectare, the additional land area required to support a horse is about 20%.

The major impact of this horse-powered system is the 10-fold increase in manpower per hectare that is required over the tractor-powered system (Tables 2 and 3). The total energy input into the horse-powered system is 7.2 × 10[6] kcal compared with 8.4 × 10[6] kcal in the tractor system (Tables 2 and 3). The resulting reduced energy input with horse power makes it slightly more energy efficient than the tractor-powered system. Although the energy ratio for a horse-powered system is 3.4 : 1 versus 2.9 : 1 for the tractor-powered system (Tables 2 and 3), at present it is not economically feasible to use this much manpower to produce corn.

Table 3. Energy inputs per hectare for corn production in the United States employing horse power.

Item	Quantity/ha	kcal/ha
Labor	120 hr	70,000
Machinery	15 kg	27,000
Horse		
Corn	136 kg	477,300
Hay	136 kg	409,000
LP Gas	80 l	616,400
Electricity	33.4 kwh	95,500
Nitrogen	151 kg	2,220,000
Phosphorus	72 kg	216,000
Potassium	84 kg	134,000
Lime	426 kg	134,400
Seeds	18 kg	445,800
Insecticides	1.4 kg	119,950
Herbicides	7.8 kg	777,500
Drying	7,000 kg	1,437,800
Transportation	150 kg	38,550
Total		7,219,200
	Output	
Total yield	7,000	24,500,000
kcal output/kcal input		3.4

ALTERNATIVES FOR REDUCING ENERGY INPUTS FOR CROP PRODUCTION

Energy from fossil fuels flowing into agricultural crop production can be reduced by altering some agricultural practices and substituting practices that require less fossil fuel energy. Most of the alternative practices are more in harmony with the natural ecosystem and thus require less manipulation of the ecosystem and subsequently less energy expenditures. Several alternative agricultural practices for reducing energy inputs in corn production are discussed below.

Soil Nutrients

When crops are harvested, significant quantities of soil nutrients are lost and must be replaced to sustain high yields. For example, when 7000 kg of corn are harvested, an estimated 104 kg of nitrogen, 19 kg of phosphorus, and 22 kg of potassium is removed from the hectare of land (Table 4). A significant

Table 4.   Loss of nutrients annually on 1 ha of land producing 7000 kg of corn and a soil erosion rate of 22 tonnes/ha.[a]

| | kg/ha of Nutrients Removed | | |
	Nitrogen	Phosphorus	Potassium
Harvest of corn	104	19	22
Soil erosion	70	20	440
Other	3	0	0

[a] Pimentel et al., 1976; Pimentel and Moran, 1983.

loss also occurs when soil is eroded by wind and rain as often happens on corn fields (Table 4). This loss in nutrients ranges from equal for phosphorus to 20 times that for potassium, which is removed by the harvest of the corn grain. Thus, a sensible alternative would be to reduce soil erosion.

In this analysis of conservation and replenishment of soil nutrients for corn production, the focus will be primarily on nitrogen because this is the most energy intensive to produce. About 14,700 kcal of fossil fuel are expended to produce 1 kg of nitrate fertilizer compared with 3000 kcal/kg for phosphorus and 1600 kcal/kg for potassium (Lockeretz, 1980). The process of manufacturing fertilizer is energy intensive.

Livestock manure can be used to substitute for some or all commercial fertilizer. Manure not only is a source of nutrients that crops need, but its addition to soil aids in reducing soil erosion and improving soil structure (Neal, 1939; Zwerman et al., 1970). Current U.S. livestock manure production is estimated to be 1.1 billion tonnes per year, with about 420 million tonnes produced in feedlots and confined-rearing situations (Miller and McCormac, 1978; Van Dyne and Gilbertson, 1978). More than 70% of this collected manure is applied to land and provides agriculture with about 8% of its nitrogen, 20% of phosphorus, and 20% of potassium. Unfortunately, about one-half of the nitrogen is lost before it reaches the intended crop because of poor manure handling and management practices (Muck, 1982).

The prime difficulty in utilizing manure as a nutrient source is that large quantities must be handled to obtain nutrients. For example, 1 tonne of cattle manure contains only 5.6 kg of nitrogen, 1.5 kg of phosphorus, and 3 kg of potassium (Pimentel et al., 1973). Then, assuming that the cropland is less than 1.5 km from the manure source, about 30,000 kcal of fuel is required to transport and spread this amount of manure using a tractor and spreader (Pimentel et al., 1982a).

Compare this to using a commercial fertilizer to supply the same quantities of nutrients as are in 1 tonne of cattle manure, then the energy expenditure for an equivalent of fertilizer would be almost three times that for manure or 91,600 kcal. Thus, utilizing livestock manure for a nutrient source can provide significant energy savings, assuming that the manure has to be transported only 1.5 km. (The break-even transport distance based on energy input versus the fertilizer value is 4.5 km.) Of course, the quantities required to fertilize 1 ha of corn are large, requiring about 27 tonnes of cattle manure to supply the necessary nitrogen. Clearly, applying 27 tonnes of manure would require more manpower than applying less than 0.3 tonne of commercial fertilizer to supply an equivalent amount of nutrients. However, the energy saving would be threefold using manure.

In the days before commercial fertilizers were so universally used in corn production, corn often was planted in rotation with a legume crop such as sweet clover (Pimentel, 1981). Planting sweet clover in the fall after corn is harvested and plowing it under one year later can add nearly 170 kg of nitrogen per hectare (Willard, 1927; Scott, 1982). However, because 2 ha of land must be cultivated to raise 1 ha of corn, widespread use of this practice is limited.

When legume rotations are not feasible, legumes can be planted between corn rows in August and then this so-called green manure is plowed under in early spring when the field is prepared for reseeding. Winter vetch and other legumes planted in this manner yield about 150 kg/ha of nitrogen (Sprague, 1936; Mitchell and Teel, 1977; Scott, 1982). Use of a cover crop also protects the soil from wind and water erosion during the winter and has the advantage of adding organic matter to the soil. Of course, the disadvantage is that the green manure must be plowed under during the spring when the farmer is pressed for time to plant the major crop.

Cultural Practices

Relying on no-tillage and minimum tillage systems would reduce the tractor fuel inputs normally required to plow and disc soil used in conventional tillage. To till 1 hectare of soil with a 50-hp tractor uses about 60 liters of diesel fuel. On the average, no-till planting uses only 15 liters of fuel. Thus, the saving might be 45 liters or 513,630 kcal of fossil fuel.

In this situation, however, herbicide and insecticide use must be increased over conventional tillage to control weeds and deal with the often increased insect and slug problems (USDA, 1975). Pesticide use is sometimes doubled with no-till, and if this is required, the energy inputs would be greater than conventional tillage when pesticides are included in the energy budget. Another energy input with no-till is the large amount of corn seed (about 13% more than conventional tillage) that is required to offset poor germination (Phillips et al., 1980).

Whether energy inputs are greater or about the same in total for no-till compared with conventional tillage will depend on the various practices used, tractor size, type of soil, abundance of pests, and environmental factors. No-till, however, does have two distinct advantages over conventional till because its use reduces manpower inputs and helps control soil erosion. Although reducing the manpower input per hectare by 2 hr will not reduce the total energy input by much, reducing soil erosion from 22 tonnes/ha to 0.2 tonne/ha would decrease nutrient losses by 1.7×10^6 kcal which is significant.

One way to minimize weeds on small "no-till" plots is to apply organic matter (leaves and similar organic matter) at a depth of about 5 cm (Pimentel, unpublished data). This organic matter will control weeds and at the same time provide ample soil nutrients. To facilitate planting, a 15 cm opening can be made in the organic matter and the seeds inserted in the uncovered soil. Once the young plants are up about 10 cm, the organic matter can be pushed back around the base of the plants.

The benefits of using organic matter include slowing soil erosion, conserving soil moisture, and minimizing tillage. Thus, this technology has several advantages in reducing energy inputs in crop production, but again the labor input is large.

Pest Control

Successful breeding of food and fiber crops with resistance against disease, insect, and weed damage substantially reduces energy inputs for pest control (Pimentel et al., 1973; Pimentel et al., 1982b). For example, planting corn after soybeans eliminates the need for insecticide use for control of the rootworm complex. At the same time, of course, this technology would reduce the diverse environmental problems associated with pesticide usage.

Also helpful in reducing energy expenditure for pest control are the nonchemical biological and cultural controls like the use of parasites and predators for insect control. Natural enemies obtain their fuel directly from the pest population. In certain crops like cotton the effective use of natural enemies in integrated pest management can reduce insecticide use by one-third and provide significant energy savings (ICAITI, 1977).

Crop Drying

Corn, like many other grain crops, is usually harvested today as grain directly in the field. This corn contains about 27% moisture and must be dried to about 13% moisture before being placed in storage. To reduce the moisture level from 27 to 13% in 7.4 kg of corn requires an input of 1520 kcal fossil energy (Pimentel, 1980).

Solar energy (wind and sunlight) could be effectively used to dry corn. The corn on the cob is placed in screened corn cribs to dry. An assessment of this solar system indicates that it requires less than one-third as much energy for corn drying compared with harvesting shelled corn and then drying with fossil energy (Hudson, 1982).

Another advantage of harvesting corn on the cob is the availability of the cobs after the corn grain is removed so they can be ground and fed to cattle or burned to provide a fuel source.

Irrigation

Water is the most serious limiting factor in crop production everywhere in the world. In arid regions, if water must be pumped from underground and applied to the crop using sprinkle irrigation, energy input for the irrigation may total 12 million kcal, an energy expenditure that averages three times greater than that for rain-fed corn (Pimentel and Burgess, 1980). The energy input for irrigation may be reduced if drip irrigation is employed when water is pumped from significant depths (Pimentel et al., 1982c).

Numerous techniques are available for conserving water and most of these are the same as those for preventing soil erosion. The following techniques will both reduce soil erosion and conserve water: (1) using organic mulches as in no-till crop culture, (2) terracing, (3) planting strip crops, and (4) planting crops on the contour.

SUMMARY

Raising corn by hand is three to four times more energy efficient based on the ratio of energy inputs for production per output of corn energy. Horse power is slightly more energy efficient than tractor power in producing corn; however, 20% more land is required plus 10 times more labor than the tractor-mechanized system.

Solar energy accounts for 88% of the energy input in corn production versus 12% for fossil energy. Of course, our concern is focused on the fossil energy inputs because fossil energy resources are being rapidly depleted.

Various alternatives exist for reducing fossil energy inputs in corn production. An appropriate combination of these alternatives might reduce fossil energy inputs by 40% without reducing corn yields; this combination of alternatives might include substituting livestock manure for commercial fertilizers, employing soil conservation practices, reducing tractor inputs, planting corn after soybeans, and harvesting corn on the cob to reduce drying inputs.

REFERENCES

Coale, A. J. (1974). The history of the human population. *Sci. Am.* **231**:40–51.

Hudson, W. J. (1982). Biomass energy and food—conflicts? (Unpublished).

ICAITI. (1977). An Environmental and Economic Study of the Consequences of Pesticide Use in Central American Cotton Production. Central America Research Institute for Industry. United Nations Environment Programme. Final Report.

Lewis, O. (1951). *Life in a Mexican Village: Tepoztlan Restudied*. University of Illinois Press, Urbana.

Lockeretz, W. (1980). Energy inputs for nitrogen, phosphorus, and potash fertilizers. In *Handbook of Energy Utilization in Agriculture*. D. Pimentel (ed.), CRC Press, Boca Raton, Florida, pp. 23–24.

Miller, R. H. and McCormac, D. E. (1978). Improving soils with organic wastes. USDA Task Force, Washington, D.C.

Mitchell, W. H. and Teel, M. R. (1977). Winter annual cover crops for no-tillage corn production. *Agron. J.* **69**:569–573.

Morrison, F. B. (1956). *Feeds and Feeding*. The Morrison Publishing Company, Ithaca, New York.

Muck, R. E. (1982). Personal communication. Dept. of Agricultural Engineering, Cornell University, Ithaca, New York.

Neal, O. R. (1939). Some concurrent and residual effects of organic matter additions on surface runoff. *Soil Sci. Soc. Am. Proc.* **4**:420–425.

Phillips, R. E., Blevins, R. L., Thomas, G. W., Frye, W. W., and Phillips, S. H. (1980). No-tillage agriculture. *Science* **208**:1108–1113.

Pimentel, D. (ed.) (1980). *Handbook of Energy Utilization in Agriculture*. CRC Press, Boca Raton, Florida.

Pimentel, D. (1981). The food-land-fuel squeeze. *Chem. Tech.* **11**:214–215.

Pimentel, D. and Burgess, M. (1980). Energy inputs in corn production. In *Handbook of Energy Utilization in Agriculture*. D. Pimentel (ed.), CRC Press, Boca Raton, Florida, pp. 67–84.

Pimentel, D. and Moran, M. A. (1983). Experimental study of silviculture and harvesting procedures at Tughill Environ. Biol. Rep. 83-3, Cornell University, Ithaca, New York.

Pimentel, D., Hurd, L. E., Bellotti, A. C., Forster, M. J., Oka, I. N., Sholes, O. D., and Whitman, R. J. (1973). Food production and the energy crisis. *Science* **182**:443–449.

Pimentel, D., Terhune, E. C., Dyson-Hudson, R., Rochereau, S., Samis, R., Smith, E., Denman, D., Reifschneider, D., and Shepard, M. (1976). Land degradation: effects on food and energy resources. *Science* **194**:149–155.

Pimentel, D., Krummel, J., Gallahan, D., Hough, J., Merrill, A., Schreiner, I., Vittum, P., Koziol, F., Back, E., Yen, D., and Fiance, S. (1978). Benefits and costs of pesticide use in U.S. food production. *BioScience* **28**:772, 778–784.

Pimentel, D., Berardi, G., and Fast, S. (1982a). Energy efficiency of farming systems. Spec. Publ. Am. Soc. Agron. (In press).

Pimentel, D., Glenister, C., Fast, S., and Gallahan, D. (1982b). Environmental risks associated with the use of biological and cultural pest controls. Final Report. NSF Grant PRA 80-00803.

Pimentel, D., Fast, S., Chao, W. K., Stuart, E., Dintzis, J., Einbender, G., Schlappi, W., Andow, D., and Broderick, K. (1982c). Water resources in food and energy production. *BioScience* **32:**861–867.

Reifsnyder, W. E. and Lull, H. W. (1965). *Radiant Energy in Relation to Forests.* Tech. Bull. No. 1344. U.S. Dept. Agr. Forest Service, Washington, D.C.

Scott, T. (1982). Personal communication. Dept. of Agronomy, Cornell University, Ithaca, New York.

Sprague, H. B. (1936). The value of winter green manure crops. *New Jersey Agr. Exp. Sta. Bull.* 609.

USDA. (1975). Minimum tillage: a preliminary technology assessment. Office of Planning and Evaluation, U.S. Department of Agriculture, Washington, D.C.

Van Dyne, D. L. and Gilbertson, C. B. (1978). Estimating U.S. livestock and poultry manure and nutrient production. USDA-ESCS Publ. No. ESCS-12, March.

Willard, C. J. (1927). An experimental study of sweet clover. Ohio Agr. Exp. Sta. Bull. 405.

Zwerman, P. J., Drielsma, A. B., Jones, G. D., Klausner, S. D., and Ellis, D. (1970). Rates of water infiltration resulting from applications of dairy manure. In *Relationship of Agriculture to Soil and Water Pollution*, Proceedings of the 1970 Cornell Agricultural Waste Management Conference. Graphics Management Corp., Washington, D.C., pp. 263–270.

Effects of Soil Erosion on Agroecosystems of the Humid United States

George W. Langdale

USDA-ARS
Southern Piedmont Conservation Research Center
Watkinsville, Georgia

Richard Lowrance

USDA-ARS
Southeast Watershed Research Laboratory
Tifton, Georgia

INTRODUCTION

Failure to control soil erosion has been associated with human suffering and near extinction of civilizations. As human population densities increase in an ecosystem, more intense land use occurs and soils begin to lose productivity as soil erosion decreases fertility, tilth, and water capture. Extensive soil erosion control practices and increased inputs of nutrients, energy, and management are usually required to maintain soil productivity.

For centuries perceived problems associated with soil erosion have been expressed in the literature. Plato, in his Critias, lamented the loss of soil in

Attica (Jowett, 1892). Ruffin (1832) expressed concerns for soil erosion during the late 1700s and early 1800s in the Middle Atlantic States and actively tried to solve these problems. Hilgard (1860) warned more than a century ago that then current practices were ruining the once productive lands of the Southeast. He compared these practices with those that caused the desolation of the once fertile Roman Campagna (Jenny, 1961). Devastation by soil erosion was called to national public attention during the 1930s by Bennett (1939), and new federal programs were initiated to combat soil erosion.

Current soil erosion rates still exceed 11.2 tons ha^{-1} annually on almost 40% of major row cropland (Larson, 1981). The 11.2 tons ha^{-1} yr^{-1} rate represents about 0.089 cm of soil per year. Soil formation from unconsolidated materials only represents about 1.12 tons ha^{-1} yr^{-1} (0.0089 cm per year). At these approximate rates we are losing 1 cm of soil in 11.8 years and gain 1 cm each 118 years. Although little is known about erosion effects on ecosystems, accelerated soil erosion has local, watershed, and biospheric effects on ecological systems.

The effects of soil erosion on soil productivity were recently labeled a top national research priority by leading scientists and engineers of the nation (Williams et al., 1981; Larson et al., 1981). This suggests that scientists need to treat soil as a distinct dynamic segment of the ecological system.

RESOURCE BASE AND SOIL EROSION STATUS

National

An estimated 567 million hectares of land in the United States are used to produce food and fiber (Larson, 1981). This includes row crop, pasture, range, and forest land. An estimated 221 million hectares of this resource is used intensively for row crop and pasturelands. Approximately 63% of the 221 million hectares is prime farmland. Prime cropland is defined by Larson (1981) as land that enjoys a combination of soil characteristics, water supply, and climate that is conducive to continuous high crop yields under proper management. Only 151.4 million hectares of land, not now in cropland, remain available with medium to high cropland conversion potential. Most of this hectarage is now in pasture and range. Currently, 1.2 million hectares of farmland are lost to nonagricultural uses each year (Larson et al., 1981). Close to 400,000 hectares of this annual loss are considered prime farmland (Larson, 1981). Current projections indicate that the nation's cropland will be fully utilized in about 20 years, and additional cropland needs will have to come from lands currently considered poor for intensive agricultural production (Larson, 1981).

Table 1.   Eroded land area of the humid east, U.S.[a]

| Land Use | Soil Erosion Class | Land Area in Each Erosion Class (million hectares) | | | | | | |
		Corn Belt	Lake	Appa-lachia	Delta	SE	NE	Total
	Mg ha^{-1}							
Cropland	11.2–22.4	7.0	1.4	1.5	3.0	1.6	0.8	15.3
	22.4	6.7	0.9	1.8	1.5	1.2	0.8	12.9
	Subtotal	13.7	2.3	3.3	4.5	2.8	1.6	28.2
Pasture	11.2–22.4	1.1	0.1	0.7	0.3	0.1	0.1	2.4
	22.4	0.9	0.04	0.8	0.3	0.04	0.1	2.2
	Subtotal	2.0	0.14	1.5	0.6	0.14	0.2	4.5
	Total	15.7	2.4	4.8	5.1	2.9	1.9	32.8[b]

[a] Van Doren, Jr. et al., 1981.
[b] Represents 27% of 1977 pasture and cropland.

Humid East

This chapter discusses erosion by water in the humid east—land east of 96° west longitude. Millions of hectares of once productive humid east land has been irreversibly lost from cropland production during the last several decades. Soil erosion exceeds the current T values on 32.8 million hectares or 27% of the total area (119 million ha) allocated to cropland and pasture (Table 1). T values are the maximum rate of soil loss that will permit sustained crop productivity and never exceed 11.2 tons ha^{-1} yr^{-1} (Wischmeier and Smith, 1978). Soil losses exceed 22.4 tons ha^{-1} yr^{-1} on 12.7% of these 119 million hectares. When these soil erosion rates are permitted, rapid destruction of soil productivity occurs.

SERIOUSNESS OF SOIL EROSION

Crop Yield Reductions

The effects of soil erosion on crop production during the first half of this century were recognized by many scientists (Williams et al., 1981; Langdale and Shrader, 1982; Pimentel et al., 1976). Available modern technologies such as germplasm and improved soil fertility practices, as well as conservation tillage and irrigation equipment, were not accessible during this period. Because of the recent increased levels of farm production, it is difficult to extrapolate earlier research results to the last half of this century. Beginning about 1950, improved fertilization and management technologies began to mask the effects of soil erosion (Williams et al., 1981; Spomer and Piest, 1981). The result was considerable complacency concerning soil erosion control during the third quarter of this century. For the most part we have ignored the long-term effects of soil erosion on both crop productivity and ecosystem structure and function. Currently, on some corn belt fields, two bushels of topsoil are lost for every bushel of corn produced (Larson et al., 1981).

An important finding surfaced from the research accomplished during the first half of this century. That is, the seriousness of soil erosion is determined by the nature of the soil profile and plant germplasm (Baver, 1950; Bradfield, 1949). This is reflected in data presented in Tables 2, 3, and 4. Corn yield reductions are drastically less on eroded deep medium-textured loess or glacial till soils than those of highly weathered medium-textured soils. Most of the severe cropland soil erosion in the humid east occurs on four soil orders (Table 3). These are the Mollisols, Alfisols, Entisols, and Ultisols (Soil Survey Staff, 1975). When topsoil is removed from these soils, corn yield reductions vary from 8 to 40% (Table 2).

Following severe soil erosion, the ability to restore corn yields appears to be dependent on subsoil texture. On the Hapludults (Ultisols) of the southeastern United States, crop yield reductions are probably biased because of the usual hostile chemical environment. Subsoil acidity problems associated with these soils limit development of deep root systems causing reduced drought tolerance and yield potential (Rios and Pearson, 1964). Other contributing factors are the crop species and the climate associated with vegetative and reproductive growth periods (Table 3). There is also some evidence

Table 2.   Estimated corn yield reductions following topsoil removal.[a]

Soil classification	Texture		Yield Reduction (%)
	Surface	Subsoil	
Deep Medium Textured Soil			
Typic Udorthents	Silt loam	Silt loam	8–30
Typic Hapludolls	Silt loam	Silt loam	8–30
Typic Hapludalfs	Silt loam	Silt loam	9
Shallow Medium Textured Soils			
Typic Hapludults	Sandy clay	Clay	40
Typic Hapludults	Clay loam	Clay	36

[a] Langdale and Shrader, 1982.

Table 3. Estimated crop yield reductions associated with severe erosion.[a]

Soil Order	Crop Yield Reduction (%)			
	Corn	Soybeans	Small Grains	Forage
Mollisols	14	15	14	12
Alfisols	21	30	18	11
Entisols	19	23	24	16
Ultisols	42	37	27	23
Average	24	26	21	16

[a] Langdale and Shrader, 1982.

that crop species specificity may be associated with production on eroded soils (Baver, 1950; Langdale and Shrader, 1982). For example, cool-season legumes perform better on eroded soils than warm-season legumes because of less water stress during low evapotranspiration periods. Crop production on moderately eroded soils appears highly dependent on soil thickness or effective rooting depth of the soil (Table 4). This suggests a soil capacity factor that will be discussed in the next section.

Water, Tilth, and Fertility Factors

Probably the most critical effect of soil erosion is the reduction of available soil water storage. Buntley (1980) hypothetically demonstrated the effects on potential crop yields of losing available soil water storage through soil erosion (Table 5). Research accomplished by Batchelder and Jones (1972) and Thomas and Cassel (1979) showed the effects of losing topsoil on crop yields. Batchelder and Jones (1972) also determined the necessity of supple-

Table 4. Effect of soil thickness on estimated corn yields.

Surface Soil Thickness (cm)	Georgia Cecil[a] scl	Kentucky Crider[b] sil	Iowa Edina[a] sil
	Corn (Mg ha^{-1})		
13	1.00	—	—
25	3.58	1.07	6.02
38	5.52	2.26	6.52
51	6.02	3.39	6.65

[a] Langdale et al., 1979a. Depth to B_{2t}.
[b] Murdock et al., 1980. Depth to fragipan-drought stress.

Table 5. Effects of soil thickness on potential available water storage and crop yield.[a]

Soil Above Fragipan (cm)	Available Water Storage Potential (cm)	Crop Yield Potential (%)[b]
61.0 Uneroded	15.2	100
50.8	12.7	88
40.6	10.2	66
30.5	7.6	50
20.3 Severely eroded	5.1	32
10.2	2.5	17

[a] Buntley, 1980.
[b] Corn and soybean yield estimates associated with potential available water storage.

mental irrigation water to restore summer annual crop yields on Ultisol subsoil.

Losses of soil organic matter through soil erosion is another important factor with respect to crop yields. Reductions in organic matter content decreases available water and nutrients. Loss of organic matter is most critical on Ultisols (Table 6). Many years of intense management may be required to restore the organic component even with adequate soil fertility and supplemental irrigation. Rosenberry et al. (1980) showed that tillage power re-

Table 6. Soil fertility of eroded soils.

Degree of Erosion	Nutrient			Organic Matter (%)	Al sat. (%)
	N (%)	P (ppm)	K (meq/100 g)		
Georgia (Ultisols)[a]					
Slight	0.065	187	—	1.7	—
Severe	0.033	157	—	1.0	—
North Carolina (Ultisols)[b]					
Unaltered	—	29	0.24	5.5	5.3
Cut area	—	4	0.11	0.8	67.2
Kentucky (Alfisols)[c]					
Slight	—	77	0.50	1.7	—
Eroded	—	34	0.31	1.5	—

[a] Langdale et al., 1979a.
[b] Phillips and Kamprath, 1973.
[c] Frye et al., 1982.

quirements increased nearly twofold on severely eroded Mollisols. Although they showed no additional tillage power requirements on eroded Entisols, the power requirement may increase more than twofold on severely eroded Ultisols. Less organic matter accompanied by undesirable soil textures serve to create poor soil tilth.

Poor soil fertility is generally associated with eroding upland cultivated lands (Langdale and Shrader, 1982). Nitrogen and phosphorus are almost universally lower on severely eroded soils than on slight to moderately eroded soils (Table 6). On some Mollisols (deep medium-textured loess), only additional supplies of nitrogen and phosphorus (occasionally Zinc) are required to restore crop productivity (Langdale and Shrader, 1982). Thin and highly weathered Ultisols occupy the opposite extreme. Excessive supplies of dolomitic limestone usually must accompany additional nitrogen, phosphorus, and potassium fertilizers because of high acidity associated with high aluminium saturation of exposed subsoils (Batchelder and Jones, 1972; Phillips and Kamprath, 1973). Because of infiltration problems associated with eroded subsoils, supplemental irrigation becomes an expedient management practice to restore productivity on exposed Ultisols (Batchelder and Jones, 1972). Irrigation is necessary to obtain adequate plant stands and provide a desirable residue-soil matrix. Unless soil erosion control measures are actively employed, nitrogen losses associated with the sediment fraction become expensive when most soils are conventionally tilled (Table 7). Phosphorus losses are much smaller and less expensive than nitrogen because of the energy needed for synthetic nitrogen fertilizer production. However, total nitrogen losses are strongly associated with sediment transport. Nitrogen losses are considerably greater when soil loss exceeds the accepted T value of 11.2 tons per hectare per year.

Table 7. Nutrient losses associated with soil erosion.

Sediment	Total Nutrient	
	TKN	Total-P
Mg ha^{-1} yr^{-1}	kg ha^{-1} yr^{-1}	
Iowa[a]		
13.7	23.4	0.7
Georgia		
6.1[b]	10.9[b]	1.8[c]

[a] Spomer, et al., 1981.
[b] Langdale, et al., 1979b.
[c] Smith, et al., 1978.

Ecosystem Effects

Erosion as a natural process helps form landscapes which, combined with biological components and the abiotic environment, make up ecosystems. Erosion is normal when the processes of material movement take place slowly and preserve the natural balance of ecosystems. Accelerated or abnormal erosion disturbs the natural balance of ecosystems (Holy, 1980). Accelerated erosion affects ecosystems both at the source of soil loss and at the point of sediment deposition at some other location in the watershed. In addition, the measures necessary to offset lost agricultural productivity and control erosion require inputs of energy, nutrients, water, and management from sources beyond the eroded field and necessitates the addition of new cropland to compensate for lost production (Larson et al., 1983). Therefore, the ecosystem effects of erosion are felt at the field level, watershed level, and regional or even global level.

An important ecosystem effect at the field level is the loss of organic matter by erosion. Due to high concentrations in the surface soil and low density, organic matter is one of the first soil constituents removed by erosion (Lucas et al., 1977). Voroney et al. (1981) found that organic carbon levels in soils decreased drastically when water erosion was included in a model of soil organic carbon dynamics. Simulations of this model were carried out for prairie soils in Saskatchewan, but a similar loss of organic matter occurs in soils of the humid east. Due to erosional organic carbon losses, the simulated Saskatchewan prairie soils did not reach a long-term equilibrium but continued to decrease over a longer time period. In many soils organic matter serves as a major source of cation exchange capacity, a source of mineralizable nitrogen, and as a sink for a significant amount of nitrogen applied in fertilizer (Gilliam et al., 1983). The loss of organic matter by erosion will directly affect nutrient cycling in field-scale agroecosystems in at least two ways: (1) loss of nutrients associated with the organic matter and (2) loss of cation exchange capacity of organic matter (and clays) and subsequent loss of nutrient retention capacity of soils. The combined effect of these two factors and the increased inputs of fertilizers often required on more eroded sites is an increase in the potential for nutrient transport by both subsurface leaching and surface erosion.

The effects of water erosion on the biological components of soil ecosystems have apparently not been well documented. These effects are certainly quite complex and involve the interaction of the loss of organic matter, clays, and nutrients and changes in the physical properties of soils. Heterotrophic microorganisms utilize simpler organic compounds in soils as an energy source for the synthesis of microbial biomass. Soil organic fractions are more susceptible to microbial attack if they are not protected by adsorption to clay particles (Parton et al., 1982). This physical protection of organic matter (Anderson, 1979) may decrease due to erosion as both clays and organic matter are removed from soils.

Erosion in field-scale agroecosystems can have effects that extend to distant locations in the watershed. Eroded soil must eventually be deposited somewhere. Areas of the Southern Piedmont, which underwent severe erosion in the late 1800s and early 1900s, once had productive bottomland soils but were left with floodplains covered by unproductive sediments and extensive anthropogenic swamps (Trimble, 1970). Erosion also has detrimental effects on lakes, especially artificial impoundments (Beasley, 1972). Knisel et al. (1982) point out that erosion control in field source areas can have counterintuitive results in downstream ecosystems. Conservation practices in fields may actually lead to higher sediment loads in streams due to increased sediment-carrying capacity of streams and rapid channel degradation to adjust to the new reduced sediment input.

As erosion from present cropland reduces the agricultural productivity of these lands, other lands are necessarily brought into production. Larson et al. (1983) estimate that if the current 168 million hectare of cropland lose productivity at the rate of 0.1% per year, then the equivalent of 16.8 million hectare of cropland would be lost over 100 years. This loss of cropland will be made up from the 51 million hectare of land with high or medium potential for conversion to cropland but which may be more susceptible to erosion. Thus, erosion on present cropland will cause the conversion to crops of lands which may have more erosion and require more expenditures per hectare for erosion control.

The loss of productive potential due to erosion also requires increased fertilizer, water, and energy inputs, which increase pressures on regional and global reserves of energy and materials. Although an analysis of the indirect effects of erosion on the depletion of nonrenewable resources such as petroleum, phosphate, and fossil water has not been attempted, these biospheric ramifications will grow as erosion effects on productivity become more severe.

FUTURE

Conservation tillage, which includes no-tillage, appears to be increasing (Table 8). This practice should maintain soil losses within accepted T values on a large percentage of the acreage within the humid east. Future energy costs may force a return to leguminous plant species for biological nitrogen fixation to support multicrop systems. Baver (1950) suggested more than 30 years ago that soil productivity could be restored on some eroded Ultisols

Table 8. Land in conservation tillage by regions, 1979.[a]

Region	Land in Conservation Tillage (%)
Northeast	18
Corn Belt	29
Lake States	18
Southeast	18
Delta	7
Appalachian	31

[a] P. Crosson, 1981.

with a species shift to leguminous-grass mixtures. However, expensive land forming accompanied with conservation practices is also essential where land destruction has occurred with gullies. Some irreversibly eroded lands are also present and future crop production is permanently lost.

Many researchers and administrators are examining with renewed attention the ominous effects of soil erosion on the anticipated world food and fiber demands. National attention is being focused on the problem, and it is attracting public concern. Additional funds are currently being allocated to soil erosion-soil productivity research, and it is appropriate that we be concerned with these problems. "Our vital land resources and environment must be protected for ourselves and future generations" (Pimentel, et al., 1976).

REFERENCES

Anderson, D. W. (1979). Processes of humus formation and transformation in soils of the Canadian Great Plains. *J. Soil Sci.* **30**:77–84.

Batchelder, A. R. and Jones, Jr., J. N. (1972). Soil management factors and growth of *Zea mays* L. on topsoil and exposed subsoil. *Agron. J.* **64**:648–652.

Baver, L. D. (1950). How serious is soil erosion. *Soil Sci. Soc. Am. Proc.* **42**:1–5.

Beasley, R. P. (1972). *Erosion and Sediment Pollution Control.* Iowa State Press, Ames.

Bennett, H. H. (1939). *Soil Conservation.* McGraw-Hill, New York.

Bradfield, R. (1949). Soil productivity and the potential food supply of the United States. In "Food," Academy of Political Science. **23**:39–50.

Buntley, G. J. (1980). Another factor affecting soil erosion and water quality. In *Southeastern Soil Erosion Control and Water Quality Workshop.* B. F. Headden (ed.), National Fertilizer Development Center, Muscle Shoals, Alabama, pp. 60–63.

Crosson, P. (1981). *Conservation Tillage and Conventional Tillage: A Comparative Assessment.* Soil Conservation Society of America, Ankeny, Iowa.

Frye, W. W., Ebelhar, S. A., Murdock, L. W., Blevins, R. L. (1982). Soil erosion effects on properties and productivity of two Kentucky soils. *Soil Sci. Soc. Am. J.* **46:**1051–1055.

Gilliam, J. W., Logan, T. J., and Broadbent, F. E. (1982). Fertilizer use in relation to the environment. *In* Fertilizer Technology and Usage. Am. Soc. Agron., Madison, Wisconsin (In Press).

Hilgard, E. W. (1860). *Report on the Geology and Agriculture of the State of Mississippi.* Jackson, Mississippi.

Holy, M. (1980). *Erosion and Environment.* Pergamon Press, Oxford.

Jenny, H. (1961). E. W. Hilgard and the birth of modern soil science. Collana Della Revista Agrochimica No. 3. Simposio Internazionale di Agrochimica. Pisa, Italy.

Jowett, B. (1892). *The Dialogues of Plato* (transl. to English), Vol. II. Random House, New York, pp. 71–84.

Knisel, W. G., Leonard, R. A., and Oswald, E. B. (1982). Nonpoint-source pollution control: a resource conservation perspective. *J. Soil Water Conserv.* **37:**196–199.

Langdale, G. W., Box, Jr., J. E., Leonard, R. A., Barnett, A. P., and Fleming, W. G. (1979a). Corn yield reduction on eroded Southern Piedmont soils. *J. Soil Water Conserv.* **34:**226–228.

Langdale, G. W., Leonard, R. A., Fleming, W. G., and Jackson, W. A. (1979b). Nitrogen and chloride movement in small upland Piedmont watersheds: II. Nitrogen and chloride transport in runoff. *J. Environ. Qual.* **8:**57–63.

Langdale, G. W. and Shrader, W. D. (1982). Soil erosion effects on soil productivity of cultivated crop land. In *Determinants of Soil Loss Tolerance,* B. L. Schmidt, R. R. Allmaras, J. V. Mannering and R. I. Papendick (eds.), Soil Science Society of America, Spec. Publ. No. 45.

Larson, W. E. (1981). Protecting the soil resource base. *J. Soil Water Conserv.* **36:**13–16.

Larson, W. E., Walsh, L. M., Stewart, B. A., and Boelter, D. H., eds., (1981). *Soil and Water Resources: Research Priorities for the Nation.* Soil Science Society of America, Madison, Wisconsin.

Larson, W. E., Pierce, F. J., and Dowdy, R. H. (1983). The threat of soil erosion to long-term crop production. *Science* **219:**458–465.

Lucas, R. E., Holtman, J. B., and Connor, L. J. (1977). Soil carbon dynamics and cropping practices. In *Agriculture and Energy.* W. Lockeretz (ed.), Academic Press, New York.

Murdock, L. W., Frye, W. W., and Blevins, R. L. (1980). Economic and production effects of soil erosion. In *Southeastern Soil Erosion Control and Water Quality Workshop.* B. F. Headden (ed.), National Fertilizer Development Center, Muscle Shoals, Alabama, pp. 31–35.

Parton, W. J., Persson, J., and Anderson, D. W. (1982). Simulation of organic matter changes in Swedish soils. In Proceedings of the Third International Conference on State of the Art in Ecological Modeling, Fort Collins, Colorado.

Phillips, J. A. and Kamprath, E. J. (1973). Soil fertility problems associated with land forming in the Coastal Plain. *J. Soil Water Conserv.* **28:**69–73.

Pimentel, D., Terhune, E. C., Dyson-Hudson, R., Rochereau, S., Samis, R., Smith, E., Denmon, D., Reifschneider, D., and Shepard, M. (1976). Land degradation: Effects on food and fiber resources. *Science* **194**:149–155.

Rios, M. A. and Pearson, R. W. (1964). The effects of some chemical environmental factors on cotton root behavior. *Soil Sci. Soc. Am. Proc.* **29**:232–235.

Rosenberry, P., Knutson, R., and Harmon, L. (1980). Predicting the effects of soil depletion from erosion. *J. Soil Water Conserv.* **35**:131–134.

Ruffin, E. (1832). An essay on calcareous manures. (J. C. Sitterson, ed., 1961). Belknap Press, Cambridge, Massachusetts.

Smith, C. N., Leonard, R. A., Langdale, G. W., and Bailey, G. W. (1978). Transport of agricultural chemicals from upland Piedmont watersheds. U.S. Environmental Protection Agency, Athens, GA and U.S. Department of Agriculture, Watkinsville, GA. Final Report on Interagency Agreement No. D6-0381. Pub. No. EPA-600/3-78-056.

Soil Survey Staff. (1975). *Soil Taxonomy.* USDA, Soil Conservation Service AH 436.

Spomer, R. G. and Piest, R. F. (1981). Soil productivity and erosion of Iowa loess soils. Paper No. 81-2054 presented at Summer Meeting of Am. Soc. Ag. Engs., Orlando, Florida, June 21–24.

Thomas, D. J. and Cassel, D. K. (1979). Land-forming Atlantic Coastal Plain soils: Crop yield relationships to soil physical and chemical properties. *J. Soil Water Conserv.* **34**:20–24.

Trimble, S. W. (1970). Culturally accelerated sedimentation on the middle Georgia Piedmont, Athens, Georgia. M. A. thesis, University of Georgia.

Van Doren, Jr., D. M., Barrows, H. L., Engelstad, O. P., Foster, G. R., Koos, E. J. J., Langdale, G. W., Lucy, R. L., Mannering, J. V., Martin, W. P., Mayon, Z. B., Orth, P. G., Osborne, G., Ryan, J. A., Seitz, W. D., Smith, Jr., A. E. (1981). Humid regions: Conservation needs and technology on agricultural land. In *Soil and Water Resources: Research Priorities for the Nation.* W. E. Larson, L. M. Walsh, B. A. Steward, and D. H. Boelter (eds.), Soil Science Society of America, Madison, Wisconsin, pp. 41–61.

Voroney, R. P., VanVeen, J. A., and Paul, E. A. (1981). Organic C dynamics in grassland soils. 2. Model validation and simulation of the long-term effects of cultivation and rainfall erosion. *Can. J. Soil Sci.* **61**:211–224.

Williams, J. R., Allarmas, R. R., Renard, K. G., Lyles, L., Moldenhauer, W. C., Langdale, G. W., Meyer, L. D., Rawls, W. J., Darby, G., and Daniels, R. (1981). Soil erosion effects on soil productivity: A research perspective. *J. Soil Water Conserv.* **36**:82–90.

Wischmeier, W. H. and Smith, D. D. (1978). Predicting rainfall erosion losses—a guide to conservation planning. U. S. Dept. Agric. Handbook No. 537. Washington, D.C.

Comparative Nutrient Cycles of Natural and Agricultural Ecosystems: A Step Toward Principles

Robert G. Woodmansee

Department of Range Sciences
Colorado State University
Fort Collins, Colorado

INTRODUCTION

Two central characteristics of natural ecosystems are their tendency to accumulate nutrients against the forces of erosion, fire, leaching, or volatilization caused by episodic events and their tendency to persist through time. The thesis of this chapter addresses these characteristics in relation to agricultural ecosystems. I believe that our ability to maintain sustainable agriculture (over decades and centuries) will be dependent largely on our successfully manipulating the interactions of biotic and abiotic factors that operate within ecosystems to produce system nutrient accumulation and persistence. Furthermore, I argue that our understanding of those interactions is dependent on our ability to produce a consistent general concept of ecosystem behavior into which specific knowledge is placed. I will attempt to present aspects of an ecological concept that can serve as a structure for integrating many of the facts and/or observations that we already know so that we can use our ability to manipulate ecosystems and simultaneously not jeopardize these two system characteristics.

Specifically, I will (1) discuss some attributes of nontilled ecosystems that cause them to maintain or accumulate nutrients and to persist through time; (2) discuss some attributes of cultivated ecosystems that cause them to be altered through time, often lose nutrients, and potentially lose productivity; (3) discuss some attributes of alternative tillage practices that more closely mimic natural ecosystems with regard to persistence, nutrient balance, and productivity; and (4) integrate some of the information from points 1–3 to formulate a set of "principles" that govern ecosystem behavior and response.

This chapter contains no citations of literature. Some of the ideas I present have been studied and reported upon ad nauseam and are not controversial; others are less traditional and may be more controversial because they have not been tested repeatedly in a large number of systems. My intention is to present some sound ecologically based concepts, as I view them, and not become mired in the level of detail that would require an extensive literature review.

CHARACTERISTICS OF PERSISTENT ECOSYSTEMS

All ecosystems have biogeochemical cycles that are composed of similar functional components and processes. A simple model of what I believe to be the minimal set of functional components (boxes) and processes (arrows) necessary to maintain ecosystem integrity and system persistence is shown in Figure 1. A guiding premise that has been recognized for decades is that all ecosystems are characterized by homeostasis among their components. Major alterations in any of the system's functional components may cause significant changes in other components, thereby altering the internal structure of the system (i.e., vegetation, community composition, organic matter dominance, animal community dynamics, etc.), but ecosystem persistence is maintained even though a new homeostatic state may be developed. If any of the functional components of Figure 1 are eliminated or reduced to such a degree that they become ineffective, the ecosystem will become vulnerable to complete collapse driven by abiotic factors, usually erosion in terrestrial environments. Some ecosystems can develop without large contributions from all of these major components, but I contend such systems are highly vulnerable to total disruption.

My focus in this section centers on some of the functional roles of the components of the model shown in Figure 1 and not their taxonomic titles. Plants may be represented at various periods of time by different taxa, but some of the general functions of any plants are to persist and to produce

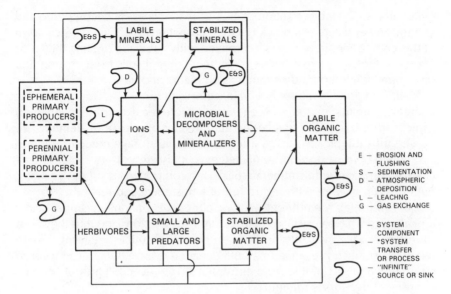

*EVERY TRANSFER IS CONTROLLED BY ONE OR MORE INTRINSIC OR EXTRINSIC ENVIRONMENTAL FACTORS.

Figure 1. Model of the components and processes that are essential for the persistent functioning of ecosystems.

reduced carbon compounds that serve as energy sources for all organisms except chemoautotrophs. Among the consequences of these functions are the production of initial carbon compounds that are or will be chemically associated with essential nutrient elements in dead organic matter and the uptake and circulation of mineral elements from soil solution and thus the reduction of the probability of those ions being lost from the ecosystem as leachate or in sediments. Plant canopies physically influence soils by reducing the energy of water and wind, thus reducing erosion. Also, plant canopies moderate soil temperature by reducing direct solar radiation and reducing reradiation.

The functions of microbial decomposers and mineralizers that enable them to persist are to alter the structure of nonliving organic matter, either reducing its complexity (decomposition) or increasing it (synthesis of secondary compounds); to cause the release of mineral ions to solution; and to take up available nutrient ions.

Labile organic matter provides a major, if not principal, portion of nutrient elements to the living components of ecosystems. These nutrients are released from the dead organic matter through the action of microbial mineralizers. The labile fraction of organic matter is easily decomposed and mineralized and acts as a quick release nutrient reserve. One physical form of this component in reality is intermixed with stabilized organic matter as plant residues (litter) and serves to protect mineral soils from erosion and moderates soil temperatures much as do plant canopies.

Stabilized organic matter is a major reserve of nutrient elements, but this fraction of organic matter is biochemically more recalcitrant than labile organic matter. It is most often larger, but the rates of release of elements are slower; thus it typically acts as a relatively stable storage compartment. In addition to being a storehouse of elements, it also serves as a major ion exchange complex within mineral soils. This component is often the major buffer within systems with regard to nutrient storage and release.

Labile minerals are nutrients that are either held on ion exchange sites or are salts. They have very rapid equilibrium with solutions and thus they are a source of readily available nutrients to plants and microorganisms. These minerals usually represent a small but dynamic reserve, which can resist or retard mechanisms of loss from ecosystems. However, because it is small, the risk of large loss is minimized.

Stabilized minerals can be salts of low solubility or primary and secondary minerals. They can represent very large pools, but nutrients stored in these forms are generally very slowly released and may contribute little to the short-term functioning of the ecosystem. In severely degraded systems or over long time frames, they may be the principal source of some nutrients.

Herbivores and small and large predators can have direct influences on ecosystem biogeochemistry, as do the heterotrophic microorganisms, by converting organically bound nutrients into mineral or simpler organic compound forms and then depositing those chemicals within the system. Possibly a much more important role of these components is that of regulation of the dynamics of plants and microbial decomposers and mineralizers. By influencing those dynamics, the rates of the processes associated with these two components may be controlled in ways to contribute to homeostasis. I readily admit the interactions of plants, microorganisms, and their consumers are not well understood. Much more research will be required to solve this puzzle.

The key issue in evaluating ecosystem persistence is not which taxa or chemical compounds perform the functions but, rather, whether all functions are being performed. In agriculture we want to exploit productivity that is useful to man, but we must guard against manipulations that jeopardize the ecosystem's functional components and thereby risk its ability to acquire and retain essential elements and its capability to persist.

COMPARISON OF NATURAL AND CULTIVATED ECOSYSTEM PROPERTIES

Abiotic Factors

Several ecosystem characteristics that are important when considering system persistence, and its converse degradation, are given in Table 1. Infiltration rates of water are typically higher in natural than in cultivated systems, while potential runoff is lower. Thus, erosion rates are lower under natural conditions because of the presence of a plant canopy most or all of the year and the presence of surface litter, debris, and rocks. Erosion can be devastating to ecosystems because it is capable of reducing or removing entire vital components (Fig. 1).

Soil water loss rates and total water loss are typically higher due to evaporation in natural systems because of more complete exploitation of the soil volume by perennial roots. Exceptions to this observation are when annual crops are fully developed—a situation that exists only for short periods of

Table 1. Characteristics of natural and cultivated ecosystems that directly influence their propensity to accumulate nutrients and persist through time.

	Natural	Cultivated
Abiotic		
Infiltration rates	High	Low
Runoff	Low	High
Erosion	Low	High
Presence of canopy	High	Low
Litter and debris	High	Low
Rocks	High	Low
Soil water loss to transpiration	High	Low
Soil colloids	High	Low
Leaching losses	Low	High
Soil temperature	Low	High
Biotic		
Internal cycling by plants	High	Low
Synchrony of plant–microorganism activity	High	Low
Temporal diversity of organism activity	High	Low
Balance of plant–microorganism activity	1	<1
Structural diversity of plants	High	Low
Genetic diversity	High	Low
Reproductive potential	High	Low

time during the year. Because more water is lost to the atmosphere, less is available to move through the soil profile carrying nutrients as leachates.

Soil organic colloids, which form a large fraction of the ion exchange sites and increase water-holding capacity, are higher in most natural systems than in comparable cultivated systems. The reason for the lower levels in cultivated systems is the oxidation and loss of organic matter commonly reported for the 60–80 years following cropping. Concomitant with the oxidation of organic matter is the mineralization of essential elements that were contained in the organic matter. Often, as a result of mineralization in the absence of growing plants, greater quantities of mobile ions are lost from cultivated systems. The biological oxidation and mineralization processes are enhanced in cultivated systems, partly because of higher soil temperatures brought about by reduced shading by canopies and higher soil water contents resulting from reduced water loss.

All of these abiotic factors interact to influence the capacity of an ecosystem to retain elements against the forces of erosion and leaching.

Biotic Factors

The biotic components of an ecosystem exert direct influence on the capacity of an ecosystem to retain and accumulate essential elements, and thus persist. Several of these influences are shown in Table 1. Both annual and perennial plants are capable of using single atoms of an essential element many times for different purposes without the atom leaving the living organism. For example, in annual plants, an atom of nitrogen may be used in an amino acid to form a protein in a root cell, remobilized as an amino acid and translocated to a leaf to be used in another protein, again remobilized and translocated to a developing seed. Another example, in perennial plants, is the use of an atom of nitrogen in a protein in a leaf one growing season, remobilization and translocation to a storage location in the plant crown or perennial roots, and subsequent remobilization and translocation back to new leaves the following growing season. The results of these reuse mechanisms are that the plant nutrient requirements from the soil system are reduced, the plant can accumulate elements, and a large portion of the biologically active elements in a system are not made vulnerable to loss from the soil system via leaching or erosion. The internal recycling of elements is more important in the success of natural ecosystems than cultivated ecosystems.

The synchrony of plant and microorganism activity is another important biotically controlled nutrient-conserving mechanism. In natural ecosystems, when water and temperature are favorable for biotic activity, plants grow, produce new carbon compounds, and take up nutrients. Simultaneously, soil microorganisms reproduce, grow and die, and perform their decomposition and mineralization functions. Most or all essential elements not taken up by new microbial cells are taken up by plants. Thus, the elements are recycled. In cultivated systems, particularly under single cropping, substantial periods exist in which growth conditions are favorable and microorganisms are active and mineralizing nutrients, but no plants exist. During these periods, the ecosystem is vulnerable to loss of essential elements through leaching or erosion.

Because organisms in ecosystems tend to grow until all available nutrients are immobilized, plant and microorganism activities tend to limit one another and thus be in balance. Plants in natural ecosystems cannot continue to grow if microorganisms do not mineralize nutrients. Microorganisms cannot mineralize nutrients if plants do not produce litter, which is a source of nutrients and energy required for microbial activity. In most natural ecosystems plants seem to have a greater capacity to take up nutrients than the microorganisms have to mineralize excesses. In single-cropped, cultivated systems microorganisms on a year-long basis have a greater capacity to mineralize nutrients than plants have to take them up simply because the plants do not exist for long periods during the year. If water is available for leaching or producing runoff, or wind is present to cause erosion, nutrients can be lost.

Plants of natural ecosystems tend to have a broad structural diversity; that is, some plants have shallow, fibrous root systems, some have deep tap roots, and some have both. These various types of rooting systems in the same ecosystem lead to the efficient exploitation of available nutrients in an entire soil volume when various water regimes are present. In cultivated systems plants tend to have a single type of rooting system, usually shallow, that effectively exploits nutrients from the surface layers of the soil but that may be ineffective in removing them from deeper depths. Those nutrients then are vulnerable to loss via leaching.

The biotic components of natural ecosystems, with a few exceptions, exhibit a broad species diversity, whereas cultivated systems generally have a very limited species diversity that is dominated by propagules of previous crops and typically common "weeds." Natural systems, when disturbed, typically have a "bank" of many endogenous propagules ("rare species") that can potentially replace predisturbance, dominant taxa. Exactly which taxa will be successful depends largely on the characteristics of the abiotic environment in which they must develop. But successful replacement will occur unless excessive physical disturbance has taken place. Following termination of long-term cultivation, biotic replacement will depend on immigration of propagules. If this immigration is slow, nutrients can be lost in large quantities and ecosystem persistence may be threatened.

ATTRIBUTES OF ALTERNATIVE TILLAGE SYSTEMS THAT MIMIC "NATURAL" ECOSYSTEMS

Cropping systems that have been developed to minimize cultivation and leave as much organic residue at the soil surface as possible are functionally more representative of natural systems than are conventionally cultivated systems. Several examples of the attributes of alternative tillage systems that mimic "natural" ecosystems are given in this section.

Reducing cultivation generally reduces soil compaction and preserves plant litter and debris at the soil surface. The combination of these two factors increases the rate of infiltration of water into the soil, thus increasing the amount available for storage. Consequently, less water is available to run off. The water that does run off is slowed by the plant litter and debris, thus reducing its energy for erosion. The litter and debris also physically protect the soil from the direct action of water.

The presence of litter and debris at the soil surface also moderates the temperature fluxes within the soil body. Consequently, temperatures during the growing season, but before full canopy closure, are generally lower than in similar bare soil situations. As a result of lower growing season soil temperatures, microbial activity rates are reduced and mineralization retarded—a desirable situation if plants are not available to take up the mineralized ions. Somewhat offsetting this temperature factor, however, is an increased availability of water, which can enhance mineralization and promote leaching.

Management of crop residues to keep them at or near the soil surface and the lack of the "stimulatory" effect of tillage seem to retard the overall net loss of soil organic matter. In fact, some evidence indicates the soil organic matter can increase where crop organic matter is thus managed and augmented by fertilizer.

The key biological property that is manipulated by this form of crop management is microbial and invertebrate activity. The reduced rate of organic matter loss is accomplished by *not* stimulating their activity. Cultivation of the residues into the soil not only leads to increased soil temperature but intermixes the organic residues with the mineral soil components, thereby creating greater surface contact of the residues with soil-inhabiting microorganisms and enhancing decomposition (CO_2 evolution) and mineralization. The slower the rate of decomposition and subsequent net mineralization, the longer those nutrients will remain bound in organic matter. If decomposition and gross mineralization are retarded so that net mineralization is synchronized with plant growth, losses of nutrients will be minimal. With the nutrients thus held within the ecosystem, not only are they available for plant growth but excess nitrogen, phosphorus, and sulfur may also be available to be incorporated into soil organic matter by microorganisms either as waste products or low energy-yielding nutrient-rich microbial organic debris. Minimum tillage agriculture thus maintains some of the attributes of natural ecosystems by regulating the degree of synchrony among microbial activity, plant growth, and element uptake.

As with cultivation agriculture, minimum tillage severely reduces plant structural diversity, biotic diversity, and reproductive potential. These attributes of natural ecosystems can be somewhat mimicked if crops are rotated or intermixed. Minimum tillage systems also require large amounts of herbicides for maintenance. The effects of these chemicals are not well known. Also, implementation of minimum tillage agriculture sometimes causes some severe pest ecology problems not common to natural ecosystems.

EXAMPLES OF ECOLOGICAL PRINCIPLES

Several attributes can be described that characterize the ability of many natural ecosystems to capture and retain nutrients, and thus persist. I will present in this section a few examples of such attributes that are evident in many ecosystems. I suggest that these attributes are so common in natural systems that they can be presented as principles of ecosystem functioning. These principles may be useful in evaluating management practices that are applied to agricultural lands. Where agricultural practices can allow mimicking of these attributes, they should be encouraged. However, factors such as disease control or economics may mitigate against their use.

Principle
Vegetation canopies, living or dead, reduce the energy of rain, running water, and wind thereby reducing erosion potential and enhancing sedimentation potential.

Where vegetation canopies exist, erosion rates are reduced and often deposition of sediments occurs. Where canopies are removed or greatly reduced for major periods of the year, erosion potential is increased and the sediment-trapping capacity of the ecosystem reduced. Thus, both nutrient capture and retention functions of the ecosystem are reduced.

Principle
Where not intimately protected by vegetation canopies, the surface layers of soils, including litter, of natural ecosystems often reduce the energy of water and wind, thereby reducing erosion. They also absorb water readily and reduce water loss by reducing evaporation.

Soils that are not directly protected by a vegetation canopy are often protected from erosion by various physical features. Prominent among these features are litter cover in forests, erosion "pavement" in arid environments, and surface crusting caused by algae or clay aggregation. Litter cover enhances infiltration of water, and all of these features reduce evaporation rates when water is stored in the soil profile. Cultivation admixes the soil material and often the "protective" features of the soil are lost.

Principle
Vegetation canopies and litter reduce soil surface temperatures during the seasons of most biotic activity thereby reducing the rates of microbial activity and evaporation.

Temperature is an important control of both microbial activity and evaporation. Vegetation canopies and litter reduce direct radiation at the soil surface. Consequently, less heat builds up in the soil body, resulting in lower soil temperatures. These lower soil temperatures result in slower rates of decomposition and mineralization. The slower rates of evaporation result in longer periods of soil water availability. The combination of the temperature and water factors tends to reduce the rate but prolong the period of microbial activity. Tillage practices that remove canopies and litter tend to cause shorter bursts of high microbial activity which, if out of phase with plant growth processes, can lead to excess mineralization and possible subsequent gaseous or leaching losses.

Principle
Natural ecosystems exhibit a tight synchrony of plant and microbial activity.

When conditions are favorable for activity of microorganisms in ecosystems, there are generally present some active plant groups that can take up mineral elements. Throughout the growing season different plant groups may perform this function. Rarely will the mineralization potential of microorganisms exceed the uptake potential of the plants. Tillage systems often create conditions where microbial activity is high but plants are not available to take up mineralized nutrients. Under these conditions, losses can occur.

Principle
Natural ecosystems typically have abundance of wide carbon-to-nitrogen ratio plant residues during periods of plant inactivity. These residues are typically in locations spatially unfavorable for rapid decomposition and mineralization.

Many ecosystems have developed in environments that typically have periods of plant inactivity because of low temperatures or soil water availability. Such environments often have brief periods of favorable growing conditions for microorganisms (i.e., small rain showers, short periods of warm weather, etc.). These brief periods are often too short to activate plant growth. Because plant residues are not admixed with the mineral soil, their rates of decomposition and mineralization are slow. Furthermore, such plant residues typically have carbon-to-nitrogen ratios that are greater than 25–30, and little net mineralization takes place because nutrients from decomposed residues are immobilized into microbial biomass. Tillage systems that accomplish the mixing of plant residues with the mineral soil hasten the decomposition process and narrow the carbon-to-nitrogen ratio, thereby promoting more rapid mineralization. If plants are not active when net mineralization commences, losses may occur.

Principle
Plants of natural ecosystems typically retain a large proportion of their nutrients within their living tissues by redistribution and reuse. This recycling renders that portion of elements unavailable for loss from the ecosystem.

This characteristic of natural perennial ecosystems is responsible for supplying a large fraction of nutrients needed for growth of new biomass and maintenance of older tissues. Furthermore, those nutrients that are immobilized in the living biomass are not vulnerable to erosion and leaching losses. Management practices that favor annual crops obviously forego this "conservation" mechanism.

Principle
Plants of natural ecosystems display a broad heterogeneity of rooting structure. This heterogeneity seems to reflect the heterogeneity of soil water distribution that is expressed in the environment. This characteristic leads to complete exploitation of soil water and reduces the amount of water available for leaching.

The plants of natural ecosystems typically have developed root systems that can effectively exploit water additions of large or small size. Management practices that utilize plants with low structural diversity not only ineffectively use added water but increase the potential for nutrient loss by leaching if water moves below the rooting zone.

CONCLUSIONS

I have attempted to integrate simple and relatively well-accepted ecological concepts into some examples that may be useful in evaluating the impact on ecosystem functioning of alternative tillage practices in agroecosystems. Some of these general concepts have been known to farmers for years, even though their articulation has not been in ecological jargon. The concepts discussed herein have been confirmed from burgeoning research results in recent years. The concepts seem to be universal and thus may constitute principles. Many other examples of concepts that may prove very useful when adequately tested could be formulated. I believe the challenge for agriculture and ecological science in the coming years will be to test current concepts or "theories" and determine if they may, in fact, serve as principles. More importantly, new concepts will need to be developed and information generated to test them. Care will have to be exercised, however, not to fall into the trap of first generating data and then hoping to develop concepts.

ACKNOWLEDGMENTS

Partial support of this paper was obtained from grants from the National Science Foundation, BSR81-05281, BSR81-14822, and BSR82-07015. Drs. C. V. Cole, W. J. Parton, E. A. Paul, and W. K. Lauenroth all took part in the formative stages of the development of ideas for this chapter. I thank all of them. I would like to thank Janet Nevin Shepherd for typing the manuscript.

Modeling Agroecosystems: Lessons from Ecology

Edward J. Rykiel, Jr.

Range Science Department
Texas A&M University
College Station, Texas

INTRODUCTION

In this chapter I discuss the role that modeling and systems analysis can play in the study of agricultural ecosystems. In one respect this could be an easy task because there is a great deal of modeling activity in agriculture. In another respect, however, it is a difficult task because there is very little activity that an ecologist would recognize as ecosystem (i.e., whole system) modeling. There are various reasons for this situation, many of which have been discussed elsewhere (Dalton, 1975).

I will address some issues that seem pertinent, based on my personal experience, in an effort not only to stimulate communication between ecological and agricultural modelers, but particularly to engage the attention of nonmodelers who may not have considered the potential contributions of modeling and systems analysis. This chapter does not deal with the "how to" of modeling and systems analysis but rather with the conceptual basis for its application to agroecosystems. My underlying thesis is that the systems approach, modeling and systems analysis, can achieve its full potential as a tool for development of unifying concepts in agriculture only if it is thoughtfully applied in an interdisciplinary framework.

It seems to me that nonmodelers, the vast majority of agricultural workers, need to give more consideration to evaluating the potential of modeling activities. The tools of modeling and systems analysis may often be ignored or abused because nonmodelers (and perhaps some modelers as well) do not understand the nature and general purposes of models. Such a lack of understanding may lead to unrealistically high expectations of the potential performance of models. Equally likely, failure to understand the purpose of a particular model can lead to rejection of a useful tool. Modelers are often the culprit because a model cannot be properly evaluated unless the purpose is clearly stated and criteria are given to judge the model's performance.

Second, the "systems approach" needs to be demythologized. There is nothing especially magical about it. Perhaps the sometimes exaggerated claims of benefit to be gained by using this approach contribute to the wariness with which it is regarded by many in agriculture. The systems approach has significant strengths but also significant weaknesses. No purpose is served by exhaustive debate of this fact. Rather, emphasis should be placed on taking advantage of the strengths and avoiding the weaknesses by substitution of more appropriate methods.

One of the strengths of the systems approach that needs to be considered in greater detail is the definition of subsystems. By formulating a problem in terms of a whole system, recognition of system substructure is enhanced. Here is where management really works. Control of a system is invariably practiced at the subsystem level and seldom by attempting to manipulate the entire system as a single entity. Subsystems are frequently identified as structural (e.g., the plant component), but the essence of the problem is to identify subsystems that embody key interactions among components. For example, we sometimes seem to forget that farmers do not grow crops; Nature grows crops. Farmers tend the crops to facilitate their growth. They attempt to act as the system controllers (managers) in order to channel the natural ecological processes into agricultural productivity. As a system manager, the farmer is the manipulator of subsystem variables and the interactions among subsystems.

Conversely, the value of investigating subsystems with the intention of understanding and controlling them can be negated by failure to consider how they fit together to form an integrated system. The producer cannot afford to ignore the whole system, but overlooking subsystem interactions is commonplace in agricultural research.

Among the consequences of the success of agricultural research is a vast accumulation of knowledge. This wealth of information in itself forces us to a holistic or systems point of view to organize and make sense of what we know. Often it seems that a holistic approach to solving problems is irrelevant until the system has been pushed to its limits. Then we become aware of interactions that suddenly appear crucial, even though they have always existed in the system. And so, where we might have considered agriculture as a man-made system controlled by humans, we need to recall that a system dependent on living components is subject to the same natural laws as any ecosystem. We have not simply imposed another level of control but have further complicated an already complex system. It should be obvious that agricultural systems tend to resist these controls and revert to more natural ecological conditions. In this sense agricultural systems are unnatural; and it is this very fact that requires us to tend to them. This is no easy task. The complexity of even simplified agricultural systems is generally underestimated, particularly when we consider that we are attempting to stabilize inherently unstable systems and system variables.

Against this background, there are a number of ways that modeling and systems analysis can contribute to the development of knowledge and principles about agricultural ecosystems, and thereby serve a unifying role. However, this potential will not be realized without an adequate understanding of the agricultural enterprise, which incorporates elements of ecology, economics, and sociology. Moreover, this potential is not now being realized because of the lack of interdisciplinary (really, nondisciplinary) cooperation and research within the agricultural community.

THE GENERAL NATURE AND PURPOSES OF MODELS

My primary intention in discussing the general nature and purposes of models is to emphasize the point that there is room in agriculture for a wide range of modeling efforts. Often, modeling efforts are ignored because they do not conform to the expectations of one discipline or another in agriculture. Just as often, specific models are dismissed without consideration of why the models were constructed or of their potential contributions to knowledge because they do not make predictions that put money in someone's pocket, or allow a manager to see precisely what must be done in a particular situation. It is patently unacceptable to say that a model is useless because it does not solve my particular problem. Perhaps if the motivations and objectives of a modeling effort were given greater thought and were more clearly stated, such objections would decline. The single item that modelers ignore most frequently and which lays them open to such criticism is the failure to specify the criteria by which another scientist can discern whether or not the model meets its objectives. A clear statement of the purpose of the model seems an indispensable first step.

The Nature of Models

Some scientists would contend that until we have a model of the system, we do not even know what we are talking about. Jeffers (1975) suggests, for example, that ". . . the true unit of transmission of concepts in the progress of science is the model itself and not the data used to construct or validate the model." Feldman and Curry (1982) express a sentiment common among modelers, "Many complex interactions are present in these biological systems which can be fully understood only through mathematical modelling."

The application of systems analysis presumes that some model exists to analyze. Yet, it is a truism of systems science that reality must be simplified before analysis is feasible. The process of determining variables relevant to a particular problem is both a boon, because the problem is made tractable; and a bane, because many other variables are omitted. A model never captures the full reality of the actual system. No model can answer all the questions about the system we might wish to ask. The first consideration, then, for understanding models is that they are simplifications of real systems. At best, a model can capture only some small part of the reality of the entire system.

Another disconcerting aspect of models is that they are not unique. There really is no single correct way to construct a model. Some ways are easier than others, and particular model formulations may be less complicated to analyze. The art of model building is to choose a formulation that leads to a relatively straightforward method of analysis. Unfortunately, there is no general agreement on a method for arriving at such a formulation. Nor is there any general agreement on what type of model to construct for a particular problem. For any given problem, a number of different models can be formulated, all of which satisfy the requirements of the problem. This characteristic of nonuniqueness leads to a third consideration about the nature of models.

Because they are nonunique simplifications of real systems, models tend to be volatile. Two modelers dealing with precisely the same problem usually develop two different models, each modeler preferring his own invention. Most models are ignored by all except the model builder and, consequently, tend to have short lives. In addition, models are easily modified and are thus targets for tinkerers. A high compliment in ecological modeling is to have someone modify your model in preference to building another one from scratch. The combination of short life and vulnerability makes models volatile. Volatility leads to inadequate testing of a model, lack of widespread experience with a particular model, and a literature rich in different models but depauperate in any models widely used and studied. While model volatility often represents a healthy state in a vigorous area of science, it can also be construed as a sign that modeling activities are unproductive for problem solving. Modeling can be fun, but does it have any utility? This question leads to a consideration of the general purposes of modeling.

Purposes of Models

A particular model has a purpose which, ideally, is clearly stated. Yet, there is relatively widespread misunderstanding of the range of purposes for which models are used. Table 1 is constructed to reflect both an increasing level of knowledge and an increasing need for quantitative results expressed as four general purposes of modeling: exploration, explanation, projection, and prediction.

The first and perhaps least understood purpose of modeling is simply to get a better grasp on the problem. I have called this *exploration* because it amounts to an effort to understand the nature of a system or problem. Often, this type of modeling is done in someone's head and may never be commit-

Table 1. General purposes of a model.

Exploration	Objectives are very often general or intuitive, usually with no specific criteria for meeting them; main aims are insight, clarification, and understanding of factors that contribute significantly to system behavior
Explanation	General objective is to understand the structural and functional relationships among components and subsystems that explain the pattern of interconnections within the system and generate system behavior; specific objectives are related to level of resolution and the level of the study: system, subsystem, and component levels
Projection	Objective is to examine the dynamic behavior of system variables at any level (i.e., component, subsystem, or system) and the effects of changes in the values of parameters or variables and their variability, events, and decisions; the patterns of behavior represented in the dynamic relations among variables are more important than actual values of the variables
Prediction	Specific objective is to estimate future values of particular system variables and/or the nature and timing of events and decisions; emphasis is on the accuracy and utility of the prediction, and the reasonableness of the explanation of the prediction

Acceptable error can only be specified on the basis of the fitness of results to fulfill objectives of the model and/or meet the needs of practical application. Criteria should be given to permit a judgment of how closely a model meets its objectives.

ted to paper. On the other hand these models may deal with difficult problems, which can only be vaguely expressed initially. The model serves to outline concepts of a system or problem and potential ways of dealing with them. This is a very basic scientific activity, and to dismiss the models resulting from it as impractical is to deny science a fundamental tool of creativity, as well as to miss the point of the model. Exploratory and explanatory models take many diverse forms within the general categories of verbal, diagrammatic, mathematical, and computer models. The most important function of this type of model is to communicate the insights generated from reflection on the issues involved.

The second general purpose is to provide a conceptual explanation of a phenomenon in an organized manner. The major goal of the model is to juxtapose the appropriate system components in an arrangement that identifies the relationships among them. Although this activity sounds simple, lack of knowledge about relations among system components and imprecise understanding of how the system works are the crux of most management problems. As a result, the relationships expressed in an explanatory model may be, and frequently are, hypothesized connections and specific functional forms. Typically, the model is an effort to depict cause–effect relations among system variables. However, there is a direct connection between the level of explanation sought and the level of resolution of the model, which affects the ability to specify cause and effect. In general, the finer the degree of resolution (i.e., the more detailed the model becomes) the better the explanatory ability. Yet, simultaneously, more poorly known relationships are exposed, and the chain of cause and effect loses its linear aspect and becomes a web of interactions. That is, the pattern of interactions becomes a network in which it is difficult or impossible to identify simple cause–effect relationships. In effect, the explanation becomes at the same time more detailed, more hypothetical, and more difficult to quantify. The most important function of explanatory models is development of these relationships. Note that the ability to explain the general behavior of a system or the nature of a phenomenon is not synonymous with the ability to predict quantitative results of the occurrence of that behavior or phenomenon (as has been amply demonstrated in both ecology and agriculture).

The third general purpose of a model is to project over time the behavior of the system. I have called this purpose *projection* rather than *prediction* because the general qualitative features of the system's behavior are the objective, not precise quantitative estimation of future values of system variables. Projection is frequently coupled with exploration and explanation in simulation models. These models are then examined for conformity to known behavior patterns and preservation of known relationships among system variables. When known behaviors and relationships are simulated with acceptable precision, confidence that the model can tell us something about future system behavior increases. The model can then be used to examine effects resulting from alterations of system parameters and variables, with the understanding that we are looking for patterns of behavior and not exact values for variables or accurate estimates of the timing of future events. The importance of this kind of information is sometimes belittled, yet this is most often the only kind of information available about the behavior of complex systems. Economic models, for example, are often projective. They suggest a pattern of behavior of economic variables under given conditions and how this pattern will change when the conditions change. Projective models are usually mathematical and/or computer models. The most important function of projective models is to examine the system's dynamic behavior patterns and those of individual system variables.

The fourth general purpose of modeling is the one perceived to be the most utilitarian, prediction. The objective here is to estimate with acceptable accuracy the actual future values of system variables and the effects on these values of various events or decisions. Although this objective is utilitarian, it seems obvious that it is also the most precarious undertaking for a modeler or anyone else. Prediction has at least two aspects that are often ignored. The first is to make an accurate prediction based on comprehensive understanding of the system. A model of this kind is usually an attempt to integrate the purposes of explanation, projection, and prediction. Models designed for predictive purposes are most often intended not only to make an accurate prediction, but also to incorporate some realistic explanatory elements. These models can be judged both on the accuracy of the prediction and the reasonableness of the explanation.

A second aspect is simply to make an accurate prediction regardless of any ability to explain how the outcome arises. This effort does not require a system's model. These models are usually statistical but can be of nearly any sort including a crystal ball. If examining the intestines of a chicken to assay the auguries predicts what you want to know, then you have found a method of prediction that satisfies your needs. The major drawback of these models is a lack of flexibility in dynamic situations. The sole basis for judging a predictive model of the second kind is whether or not the results are accurate and achieve the pragmatic purpose (i.e., the utilitarian objective is served). The most important function of predictive models is to make an accurate prediction of future values of system variables. Needless to say, predictive modeling is a most difficult task in the world of agroecosystems.

THE SYSTEMS APPROACH

The "systems" approach to problem solving is being used fairly widely in agriculture. The implications of this particular approach are that there is something unique about a system (as opposed to a handful of components) and that systems analysis is a key element. Jeffers (1978) has taken the trouble to point out to ecologists something of the nature of systems analysis:

> Contrary to the belief of many ecologists, systems analysis is not a mathematical technique, nor even a group of mathematical techniques. It is a broad research strategy . . . it provides a framework of thought designed to help decision-makers to choose a desirable course of action, or to predict the outcome of one or more courses of action . . . systems analysis is the orderly and logical organization of data and information into models, followed by rigorous testing and exploration . . . necessary for their validation and improvement.

Systems analysis as a profession is concerned with decisions and efficiency. In discussing operations research (OR), a principal source of those mathematical techniques used in systems analysis, Wagner (1975) says that OR is,

> . . . a scientific approach to problem-solving for executive management . . . involves: constructing mathematical, economic, and statistical descriptions or models of decision and control problems to treat situations of complexity and uncertainty. . . analyzing the relationships that determine the probable future consequences of decision choices, and devising appropriate measures of effectiveness in order to evaluate the relative merit of alternative actions.

He suggests that OR should be renamed decision analysis. As if to emphasize Wagner's point, Arnold and Bennett (1975), discussing optimization, suggest that:

> There seem to be three reasons for optimizing: (i) to surround a conjectural answer with so much mathematical garbage that skeptics are afraid to challenge it; (ii) to satisfy the model builder that he has an important message; (iii) to help answer questions posed by decision-makers, which are aimed at improving their management, and so improve their welfare.

Actually, both Jeffers and Wagner take a somewhat narrow view (incidentally, the same view as Webster's dictionary) of systems analysis, preferring to consider it as something akin to an academic discipline. Most ecologists and agriculturists have taken the meaning of systems analysis in the broad context of any analysis of a system regardless of whether or not decision making and efficiency are involved; in other words as a framework for thought as suggested by Jeffers. Ecologists have particularly avoided the decision-making implications but have managed to find many useful applications of systems analysis techniques. Nevertheless, it is well to keep in mind that methods developed to deal with the problems that crop up in one disciplinary field are not necessarily suited by virtue of their success in that field for fruitful application in another. The task in this particular area is not indiscriminant plundering of methodology, but careful selection, application, and testing of methodology.

The second element of the systems approach is often referred to as systems theory. There is no real definition of the meaning of this terminology except that it deals with systems. Usually, the implication is that elements of General Systems theories (e.g., Klir, 1969) and/or elements or techniques of systems analysis are included. Systems theory does not seem to have been widely applied in agriculture, probably because there appears to be a general inability to distinguish between systems theory in the general systems sense and development of theory that leads to improved methods of analysis in the operations research sense. The two are not synonymous, although they overlap to some extent. Failure to appreciate this distinction is sometimes the basis for unfair criticism of modeling and systems analysis work.

There is a range of opinions on the actual and potential effectiveness of applying the systems approach, systems theory, systems analysis, and modeling to agriculture. Charlton and Street (1975) charge that:

> It is principally only by non-practitioners, that is, by academics and advisors, that the need to consider complete systems, rather than just their component parts, has often been neglected.

> It is unfortunate that this rather simple, naive concept of needing to look at the whole system has been developed by theoreticians far beyond the level at which the approach can still be of any practical relevance.

They suggest that the only use of systems theory in agriculture is to communicate general concepts to those unfamiliar with a particular system because:

> To anyone familiar at the practical level . . . with all the factors that can distort the real behavior of such systems, such modelling efforts are extremely simplistic and naive. They cannot be used in any way as predictive models capable of giving guidance on specific issues, or of producing information directly useable by decision-makers.

Expressing another opinion, Dent (1975) decries the lack of interdisciplinary efforts to apply systems theory. The narrowness and specialization of agricultural disciplines is Dent's culprit:

> The present lack of vertical integration in systems research in agriculture is a totally unsatisfactory condition and undoubtedly is a symptom of the intradisciplinary bound nature of the personnel involved. The real world does not come to us in disciplinary packages and the fact that researchers are locked within disciplines . . . could be a contributory factor in the general lack of impact of systems work in agriculture.

Dent's point is reinforced by Feldman and Curry (1982), who describe some of the successes and problems of applying operations research techniques in agriculture: "The interdisciplinary feature is vital in the pest management field where models must be developed in cooperation with an experimental program if results are to be put in practice." They emphasize that biological insight can help to simplify an otherwise intractable optimization problem.

Although arguments about the role of modeling, systems analysis, and systems theory are likely to persist, there is a great deal of modeling going on in agricultural research. Two journals are specifically devoted to agroecosystems (*Agricultural Systems*, and *Agriculture, Ecosystems and Environment*). Van Dyne and Abramsky (1975), for example, reviewed 127 agricultural models in one article! Modeling and systems analysis are not being slighted in agriculture.

However, it seems that these activities are not serving the interdisciplinary integrative function that they might. Often these models are special-purpose tools that provide insight into particular agricultural problems but little perspective on how the whole system is affected or operates. For example, I have relied on Dalton (1975) because it is one of the few attempts to explore and develop the systems approach to agriculture. Many of the conclusions of Van Dyne and Abramsky (1975) are relevant to this point, but four stand out: (1) models are often inadequately described; (2) models are often simply built and reported with no detailed analysis; (3) objectives of models seem to be written last, after the model is built; (4) modelers seldom refer to the potential variance (statistical properties) of values generated by the model.

On the one hand, grand scale systems models as advocated by some holistic-thinking observers are simply unsatisfactory at the level of a farm or ranch operation where day-to-day decisions are being made, or at the level of agricultural research where a herbicide is being tested for "efficacy." On the other hand, low-level, high-resolution models (of photosynthesis, for example) do not provide information necessary to make system-level decisions, although they give us insight into basic processes affecting agricultural productivity. It seems there ought to be some middle ground where useful models can be constructed that are neither too detailed nor too simplified.

Perhaps, one of the reasons that appropriately scaled models are lacking or are difficult to develop in agriculture is contained in the basic concept of "system." This concept is far too general to permit unequivocal identification of a system. Any group of components that can be connected together will constitute a system. We have not addressed the basic issue of system completeness. How do you know when you have the right components organized in the proper fashion? What is the system? In ecology, where the only unequivocally identifiable ecosystem is the biosphere, arbitrary units (often plant "communities") are defined to be the system. The most important element of ecosystem definition is whether or not the defined boundaries permit a reasonable interpretation of the inputs and outputs of the ecosystem.

We can just as well ask what is an agroecosystem. We cannot pretend to be considering a whole agroecosystem if we have not identified its components, their pattern of interconnection, their functional relationships, and the system boundaries and fluxes across these boundaries. Researchers, invoking models to illustrate concepts or suggest probable system behaviors, frequently fail to define the system they are talking about. There is much loose talk about how "the system" behaves or operates when "the system" has not been identified, defined, or otherwise described. Another point is that once the system is defined, the surroundings or the environment of that system are also defined, at least in general. A basic tenet of ecology is that the system and its environment interact to produce the observed system behavior. This is not a trivial point. This interaction often defines the nature of the inputs and outputs of the system.

Presently, there is no generally accepted classification of agricultural systems. Spedding (1975a, 1979) has suggested that efforts to model these ecosystems are inhibited by lack of such a classification because we are unable to judge how widely applicable a model of a given situation may be. He has developed at least one classification. Perhaps modeling efforts should refer to a scheme such as this to identify the scope of potential applicability. This device would overcome in part the defect of failure to define the system and begin to remedy another defect in agricultural modeling: the lack of a reasonably defined system level.

A general structure, conceived as an hierarchy, exists in ecology to identify levels of organization (this is not the same as saying that ecological systems actually are hierarchies). The generally recognized system level is the ecosystem. Within the ecosystem, various subsystems can be defined, all of which contribute to the functioning of the ecosystem. There is no comparably defined system level in agriculture with the possible exception of individual farms and ranches. This situation is analogous to defining every ecosystem, and every farm and ranch, as a unique entity. So long as this viewpoint persists, we are not likely to get far in modeling of agroecosystems. Modeling must pursue both features common to all particular systems in a class and features unique to particular situations. In agriculture we are really dealing with "supersystems" composed of ecological, economic, and sociologic systems.

SUBSYSTEMS

As far as we know, all complex systems have a structure that consists of interlinked subsystems. A modular or subsystem structure makes the components of one subsystem relatively independent of other system components and, therefore, more amenable to control. Much of agricultural research has been based on the study and control of subsystems. Precisely because control is exercised at the subsystem level of organization, this research has been very successful, and this very success has masked an important point.

Subsystem components are relatively independent because they are most directly connected among themselves. However, they are not independent of the entire system because the connections among subsystems couple all system components together. That is, a component in one subsystem is indirectly connected to a component in another subsystem by virtue of subsystem interconnections. These indirect connections become increasingly important as the intensity of management rises, yet are often dismissed as not relevant to a particular problem. This is true to some extent, but taken to the extreme, is disastrous. In fact, ignoring these indirect connections has led to the problem of desertification in many parts of the world, including the United States (Sheridan, 1981). Subsystems do not stand alone but exist and function because of the indirect connections that constitute the substructure of a system. Thus, the importance of studying subsystems is based on the need to exercise some control over the system; simultaneously, the importance of studying the whole system is based on the need to understand how the subsystems are interconnected and how control (management) of one subsystem affects all others. These are, or should be, complementary activities.

There are no complete ecosystem models in the sense of a model explicitly including all species' populations, all the ecological interactions in the system, and a level of resolution sufficient to answer any question. Models purporting to represent the ecosystem tend to be highly aggregated, and conceptually generalized in terms of ecological processes. Typical ecological subsystems might include producers, herbivores, and decomposers, all of which are aggregates of many populations. Such subsystems are frequently used in ecological systems modeling. The major drawback to this approach from the viewpoint of agriculture is that farmers and ranchers cannot make a living selling "herbivores."

Ecological aggregation results in the loss of exactly the specificity needed for farm or ranch management. There is no reason to expect that whole system agroecosystem models will be any better. However, the farmer and rancher are not qualified to determine if such modeling activity should occur in the field of agricultural research. Models employing these aggregated components may represent one means of advancing the integration of ecological and agricultural theory. They clearly have a different purpose than the prediction of how many eggs will be available for sale next Thursday. For the immediate future the most likely improvement in agricultural modeling will come from integration of subsystem models guided by the need to answer specific questions or to better understand specific interactions.

Identification of subsystems is such a long-standing activity that specific subsystems now seem entrenched. Agriculture largely follows the concept of populations as subsystems. Examples might include a wheat field or a herd of cows. The nonliving world is conveniently divided into climate, soils, and water (and, of course, the myriad ramifications of these). This organization arises from a hierarchical approach to ecological classification that proceeds from individuals to populations to communities and ecosystems. While this approach is very useful for teaching purposes, it is not necessarily appropriate for modeling and systems analysis.

A common goal of subsystem definition in both ecology and agriculture arises from the need to identify subsystems that embody interactions (relationships) (e.g., Gilbert et al., 1976) between or among populations. The subsystem of interest in grazing, for example, is the plant–animal interaction. A natural subsystem consists of cows and the particular species of plants they eat. As another example, in forestry the interaction between pine trees and pine bark beetles is one key subsystem. There is an obvious relation between pine beetles and pine trees that does not extend to pine beetles and oak trees. The beetles and pines form a natural subsystem that is important from a management viewpoint as well as an ecological viewpoint. The ecological fact is that most living things ignore one another if possible; direct connections hold subsystems together; indirect connections hold the system together.

A method for identifying agricultural subsystems is provided by Spedding (1975a). His method also illustrates a major difference with ecology. In agriculture it is possible to provide a central focus such as profit, while in ecology no such central focus is generally agreed upon. Nevertheless, this method could be adapted to ecological or other studies by identifying an appropriate focus for a specific problem. Spedding's circle diagrams are illustrated in Figures 1 and 2. The essential features of the method are that a holistic approach to agricultural systems is involved, while at the same time identifying the specific connections among components and subsystems that can be studied independently. If modeling has the potential to unify agricultural research efforts and serve as a mechanism for the development of new ideas and more effective practices, this is exactly the kind of effort that is needed.

MODELING AS A UNIFYING PROCEDURE

Table 2 lists some ways that modeling can serve a unifying function in the study of agricultural systems. Many of these are obvious upon reflection, so I limit my comments to a few points. Conceptualization of what the system is

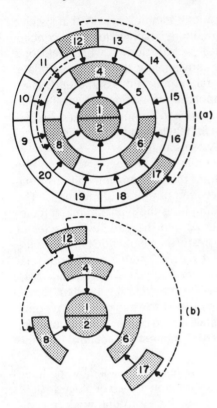

Figure 1. Subsystem definition with a circle diagram model. In this illustration, the central focus, profit, is split into expenditures and income, or more generally, input and output (items 1 and 2). The arrows indicate connections among system components. Shading identifies subsystem components. (a) Represents a model of the system; (b) represents extraction of a subsystem within this model. Note that connections among system components make it possible to: (1) identify the subsystem and (2) see how the subsystem affects the center, or objective. Also note that other factors affect the subsystem; for example, factors 9 and 20 affect factor 8 of the subsystem. (From Spedding 1979 courtesy of Applied Science Publishers Ltd.)

may be the single most important use at this time because there is a definite need to clarify the relations between components and subsystems. Definition may be as important because there is wide disparity between different agricultural disciplines both in terminology and in perception of systems and subsystems. Simulation can serve an important function because it is often impossible to conduct an appropriate experiment in the field (especially when someone's livelihood is involved), to replicate an experiment, or to test the effects of varying system parameters in a large-scale experiment. Experimentation refers to the use of a model to suggest the experiments that might be most fruitful given limited time and funds.

However, this potential has not been realized and is not likely to be if modeling of agricultural ecosystems is left strictly to modelers or ecologists or agriculturalists. The broad range of expertise required mandates interdisciplinary efforts. In my experience interdisciplinary research works best when bottom-up and top-down approaches are effectively combined. That is, when motivation for the research comes from individual scientists, who

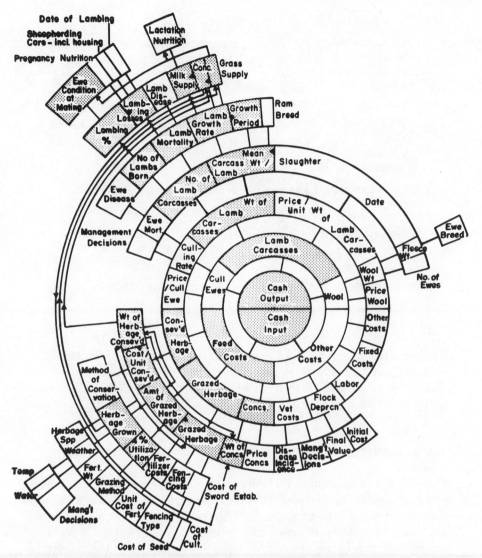

Figure 2. Application of circle diagram model. This model is based on the concepts developed in Figure 1 but is a realistic application of the method. The objective is to investigate the effect of the amount of herbage grown in a sheep production system on the profit. The subsystem of interest is indicated by shaded components and the connections among them. Note particularly that factors other than economics are included. (From Spedding 1975b courtesy of Applied Science Publishers Ltd.)

171

Table 2.   Systems modeling as a unifying procedure.

1. Formulation: development of ideas, problem statements, and approaches to solutions
    Conceptualization
    Definition
    Design

2. Clarity: formal expression of objectives, model structure, functional relations, etc.
    Organization
    Documentation
    Accounting

3. Analysis: interpretation and explanation of model and system behavior
    Statistics
    Simulation
    Systems Analysis

4. Application: use of results of analysis and of models
    Communication
    Experimentation
    Technology Transfer

recognize the need and band together to conduct the work (bottom up). And, support comes from administrators who are willing and able to promote these efforts and cut through whatever bureaucratic tangles may tend to inhibit interdisciplinary efforts (top down).

THE AGRICULTURAL ENTERPRISE

The practice of agriculture includes equal parts of ecological, economic, and sociological components (UNESCO, 1981). Failure to understand this point leads to distorted perceptions of agroecosystems and how they function. Decision making in agriculture is based on all these factors. What is ecologically impossible in one type of system may be possible in another where economics justifies the massive infusion of energy and material resources.

Often overlooked, however, is that ecological possibilities may exist only so long as the flow of resources can be maintained. A case in point is the depletion of groundwater supplies that are necessary to maintain certain types of agricultural systems in areas where they could not otherwise exist. Once the groundwater is depleted, there is no alternative but to abandon the previous system because it is ecologically impossible to maintain it. Nevertheless, the agricultural system dependent on groundwater will persist, so long as there is an economic incentive to do so, right up to the bitter end. The ecological reality is often ignored until a catastrophe occurs. To an ecologist who does not understand the nature of agriculture, particularly in developed countries, these economic decisions can appear ecologically irrational and self-defeating. To the agriculturist, on the other hand, ecologically reasonable decisions can appear economically illogical and even disastrous. The effort to design sustainable agricultural systems surely requires some resolution of these conflicts.

As an overgeneralization we might say that where food for survival is the primary concern, agricultural systems have developed that are ecologically compatible with the environment. However, these systems may not be able to support large populations. Where economic return is the primary concern, agricultural systems have developed that are often ecologically incompatible with the environment, but due to energy and material subsidies are very productive. Regardless of the kind of system, however, the producer must at least support himself. In developed agricultural systems, especially where there is specialization in crop production, the ability of the producer to support himself and others may be more a matter of economics than ecology. Consequently, it is not solely the ability to produce food that regulates the system but also the ability to make a profit from what is produced.

Agriculture is a gamble. No biological systems are totally controllable by man. We cannot prevent drought or make it rain whenever we wish. Production will vary from year to year and may fail entirely in some years. This is a part of agricultural ecology (Cox and Atkins, 1979). The best systems will produce something even in bad years, though the supply may be low. Poorly designed systems will produce nothing in bad years. A significant part of human history revolves about efforts to avoid famine by smoothing out the peaks and valleys in agricultural production.

Agriculture is a way of life. As a preferred career alternative, cultural influences may dictate that the enterprise continue even though there is no net economic profit. Many farm and ranch operations are passed from one generation to the next, engendering strong family and community ties to agricultural land holdings. Often the motivation of leaving an inheritance to succeeding generations fosters a greater sensitivity to conservation of basic resources.

Society at large is involved in agriculture. Agriculture is not something just farmers and ranchers do. Consumers can dictate to a large extent the variety and quality of agricultural products. But more importantly, consumer choices affect the ecological decisions a producer makes by affecting his economic situation. In addition, various agricultural systems have evolved because of the interaction between what is ecologically feasible and the needs and expectations of the society.

Management is the focal point of agricultural activities. Unfortunately, a biased perception exists that management means business administration and economics. Management is much more than that because it is the manager, or decison maker, who must digest and integrate information from all sources in order to make an appropriate decision. This information is no less ecological than economic. Typically, a natural resources manager does not have an adequate understanding of business administration and economics, while a business manager does not have an adequate understanding of science, particularly ecology.

These brief considerations should lead us to understand that agricultural systems are more complex than natural ecosystems because of their special human character. More complex, also, in the sense that we are far more intimately involved with these systems, as designers, producers, and chief beneficiaries. Agricultural ecosystems have all the characteristics of natural ecosystems plus the confounding factors of our efforts to manage these systems for our own purposes. Natural ecosystems will persist without us; agricultural ecosystems will not.

KEYS TO REALIZING THE POTENTIAL OF MODELING IN AGRICULTURE

1. Understanding of the holistic nature of agricultural systems derived from their ecological foundations; formulation of agricultural research in terms of systems, rather than commodities and similar political realities.
2. Appreciation of the varied roles that modeling can play; understanding the nature and purposes of models. The purpose and objectives of a model must be clearly stated and criteria given to judge model performance.
3. Interdisciplinary enthusiasm and cooperation, which have been lacking in much of agricultural research; particularly, the need for effective institutional structures combining bottom-up and top-down approaches to promote and support interdisciplinary research.
4. Recognition of the strengths and weaknesses of the systems approach:

    (a) The system concept by itself is too general to be used effectively in the context of agricultural ecosystems. What is an agricultural ecosystem? The concept must be qualified; the system must be defined.
    (b) There are no general levels of resolution that can be used consist-

ently. The level of resolution depends on the questions a model is expected to answer.

(c) Ability to deal with complicated mathematical formulations is a limiting factor. This causes the system to be bent, and perhaps broken, to fit the mathematical capability. A mathematical or mechanistic "solution" is often not necessary, but when available is the best analytical solution.

(d) A communicable concept is often more important than a detailed model or uncertain systems analysis. Systems analysis is difficult or impossible to communicate to a potential user or decision maker.

(e) No complete predictive models or systems analyses of ecosystems exist or are foreseeable, so there is no reason to expect or advocate complete predictive models or systems analyses of agricultural systems.

5. Recognition of the substructure of the ecosystem:

(a) The ecosystem is organized as a set of interconnected subsystems. This structure allows for better control of the system by permitting control of individual subsystems.

(b) The interconnections among subsystems cannot safely be ignored. The system is held together by these connections, which are required inputs and outputs of the subsystems.

(c) The ability to construct realistic subsystems is limited by our ecological, economic, and sociological knowledge. But, in principle, subsystems can be delineated by the magnitude and frequency of interactions among a set of components and by relationships that specify direct connections among components.

(d) Food chains in agricultural systems are shorter than typical ecosystems to achieve their purpose effectively. The complexity of agricultural systems arises as much or more so from socioeconomic and cultural considerations as from ecological considerations.

(e) Subsystems should embody important interactions/relations both among components and between components and the system environment, not artificially isolated groups of components.

6. Recognition that it is not necessary to have complete understanding of a system in order to partially control it and that complete control is a totally unrealistic objective:

(a) Some desirable features can be induced or obtained without having complete control of the system.

(b) Complete control of an ecological system is not possible (with the probable exception of complete destruction, and certainly with the exception of making arable land unfit for agriculture).

(c) Partial control can be obtained in the short run by practices designed to influence only specific subsystems; but this procedure, which is the common agricultural approach, fails in the long run. The degree of control can be increased by retaining flexibility in management and by adhering more closely to a broad "supersystem" point of view in research, modeling, experimentation, and application. A specific modeling problem does not have to deal directly with subsystem interactions, but it cannot simply ignore them if long-term answers are expected.

7. Understanding of the complexity of agricultural "supersystems," which are woven from the fabric of ecological, economic, and sociological systems.

ACKNOWLEDGMENTS

I would like to thank Dr. Charles J. Scifres and Dr. R. N. Coulson for reviewing a draft of the manuscript and providing cogent and stimulating comments. They are not to be blamed for shortcomings of the final document. My secretary, Chris Saltsman, did her usual spectacular job under pressure.

REFERENCES

Arnold, G. W. and Bennett, D. (1975). The Problem of Finding an Optimum Solution. In *Study of Agricultural Systems*. G. E. Dalton (ed.), Applied Science Publishers Ltd., London, pp. 129–173.

Charlton, P. J. and Street, P. R. (1975). The Practical Application of Bio-economic Models. In *Study of Agricultural Systems*. G. E. Dalton (ed.), Applied Science Publishers Ltd., London, pp. 235–265.

Cox, G. W. and Atkins, M. D. (1979). *Agricultural Ecology*. W. H. Freeman and Company, San Francisco.

Dalton, G. E., ed. (1975). *Study of Agricultural Systems*. Applied Science Publishers Ltd., London.

Dent, J. B. (1975). The Application of Systems Theory in Agriculture. In *Study of Agricultural Systems*. G. E. Dalton (ed.), Applied Science Publishers Ltd., London, pp. 107–127.

Feldman, R. M. and Curry, G. L. (1982). Operations Research for Agricultural Pest Management. *Operations Res.* **30**(4):601–618.

Gilbert, N., Gutierrez, A. P., Frazer, B. D., and Jones, R. E. (1976). *Ecological Relationships*. W. H. Freeman and Company, San Francisco.

Jeffers, J. N. R. (1978). *An Introduction to Systems Analysis: With Ecological Applications*. University Park Press, Baltimore.

Jeffers, J. N. R. (1975). Constraints and Limitations of Data Sources for Systems Models. In *Study of Agricultural Systems*. G. E. Dalton (ed.), Applied Science Publishers Ltd., London, pp. 175–186.

Klir, G. J. (1969). *An Approach to General Systems Theory*. Van Nostrand Reinhold Company, New York.

Sheridan, D. (1981). Desertification of the United States. Council on Environmental Quality. U.S. Government Printing Office, Washington, D.C.

Spedding, C. R. W. (1979). *An Introduction to Agricultural Systems*. Applied Science Publishers Ltd., London.

Spedding, C. R. W. (1975a). *The Biology of Agricultural Systems*. Academic Press, London.

Spedding, C. R. W. (1975b). The Study of Agricultural Systems. In *Study of Agricultural Systems*. G. E. Dalton (ed.), Applied Science Publishers Ltd., London, pp. 3–19.

UNESCO. Man in the Biosphere Program. (1981). Interactions Between Ecological, Economical and Social Systems in Regions of Intensive Agriculture. German National Committee MAB - Part 7, Bonn.

Van Dyne, G. M. and Abramsky, Z. (1975). Agricultural Systems Models and Modelling: An Overview. In *Study of Agricultural Systems*. G. E. Dalton (ed.), Applied Science Publishers Ltd., London, pp. 23–106.

Wagner, H. M. (1975). *Principles of Operations Research*, 2nd ed. Prentice-Hall, Englewood Cliffs, New Jersey.

ADDITIONAL REFERENCE MATERIAL

Altieri, M. A. and Whitcomb, W. H. (1979). Manipulation of insect populations through seasonal disturbance of weed communities. *Protection Ecol.* **1:** 185–202.

Bormann, F. H. and Likens, G. E. (1979). *Pattern and Process in a Forested Ecosystem*. Springer-Verlag, New York.

Breymeyer, A. I. and Van Dyne, G. M., eds. (1980). *Grasslands, Systems Analysis and Man*. Cambridge University Press, Cambridge, England. International Biological Programme 19.

Cartwright, T. C. (1979). The use of systems analysis in animal science with emphasis on animal breeding. *J. Anim. Sci.* **49**(3):817–825.

Clark, L. R., Kitching, R. L., and Geier, P. W. (1979). On the scope and value of ecology. *Protection Ecology* **1:**223–243.

Curry, G. L., Sharpe, P. J. H., DeMichele, D. W., and Cate, J. R. (1980). Towards a management model of the cotton-bollweevil ecosystem. *J. Environ. Manage.* **11:**17–223.

Feddes, R. A., Kowalik, P. A., and Zaradny, H. (1978). *Simulation of Field Water Use and Crop Yield*. John Wiley & Sons, New York.

Flint, M. L. and van den Bosch, R. (1981). *Introduction to Integrated Pest Management*. Plenum Press, New York.

Frissel, M. J. (ed.) (1978). *Cycling of Mineral Nutrients in Agricultural Ecosystems*. Elsevier Scientific Publishing Company, Amsterdam.

Huffaker, C. B. and Messenger, P. S. (eds.) (1976). *Theory and Practice of Biological Control*. Academic Press, New York.

Kirk, F. G. (1973). *Total Systems Development for Information Systems*. John Wiley & Sons, New York.

Lewis, J. K. (1969). Range management viewed in the ecosystem framework. In *The Ecosystem Concept in Natural Resource Management*. G. M. Van Dyne (ed.), Academic Press, New York, pp. 97–187.

May, R. M. (ed.) (1981). *Theoretical Ecology, Principles and Applications*. 2nd ed. Sinauer Associates, Inc., Publishers, Sunderland, Massachusetts.

May, R. M. (1977). Thresholds and breakpoints in ecosystems with a multiplicity of stable states. *Nature* **269**:471–477.

Pantell, R. H. (1976). *Techniques of Environmental Systems Analysis*. John Wiley & Sons, New York.

Seatz, L. F. (ed.) (1977). Ecology and Agricultural Production. Proceedings of a Symposium, University of Tennessee/Knoxville, July 10–17, 1973.

Smith, G. M. and Harrison, V. L. (1978). The future of livestock systems analysis. *J. Anim. Sci.* **46**(3):807–811.

Yaron, D. and Tapiero, C. S. (eds.) (1980). *Operations Research in Agriculture and Water Resources. Proceedings of the ORAGWA International Conference Held in Jerusalem, November 25–29, 1979*. North-Holland Publishing Company, Amsterdam.

Agricultural Systems and the Role of Modeling

C.R.W. Spedding

Department of Agriculture and Horticulture,
University of Reading
Reading; United Kingdom

Agricultural systems are essentially economic in nature, concerned with the efficient conversion of resources into products that are wanted by the producer or someone else who is prepared to pay for them. They are based on biological processes, but they are operated by people for a multiplicity of purposes, generally involving the export out of the system of substantial quantities of output. Agricultural systems are therefore purposive and, although embedded in an environment which may well be a natural ecosystem, their structures reflect their purposes.

The study of agricultural systems may be undertaken for a number of different purposes, but these are normally related to the purposes of the systems themselves. It is doubtful if understanding is ever sought except for some purpose but, in this case, it is certainly always so.

Essentially, study of agricultural systems is aimed at helping in (1) the operation of systems, (2) their repair, or (3) their improvement.

THE OPERATION OF SYSTEMS

Agricultural systems differ from natural ecosystems in that they do not simply function as a result of internal checks and balances. They are managed or operated by a manager who is not part of the system or, if he is, he is

179

unique among components in terms of the extent to which he influences the system toward the achievement of his objectives.

Most commonly, the manager is outside the system boundary and seeks to use, even exploit, the system for his own purposes. It is not simply that feedbacks operate through the manager because he is also subject to external influences at the point of decision.

Study aimed at helping the operator needs to identify these points of decision, since they are the "levers" by which the manager operates, and to identify the information required in order to make decisions. Channels of essential information can then be improved, and this may have to include the collection of the necessary data. It is here that computers are finding an increasing role in farm management, dealing with the processing, integration, and presentation of the data collected.

The operation of a system cannot be undertaken without an adequate picture of how the system operates, what is in it and what is not, where its boundary lies, what its main inputs and outputs are, and how these are related via component processes. Since the best picture will contain the essentials and will not be complicated by nonessentials, it is essentially a model. But it is a model that reflects the purposes of the system and its manager, and its main function is to allow the prediction of system achievement as a result of managerial decisions. This may not be enough, of course, since uncontrolled variables, notably the weather, will alter system behavior and thus, in turn, the decisions that have to be made, as well as the results of having made them.

Although such models may be relatively simple and not very detailed, they may nevertheless be too complex for the unaided human mind and thus require computers (Brockington, 1979). Where mathematical formulation is possible (and it frequently is not in the case of subsistence farming, for example), it enables preliminary experimentation on the model in ways that would be very costly or even impossible on the real system. Good examples of this are (1) computer simulation of the last 10 years weather data rather than carrying out physical experiments over the next 10 years real weather (more costly and incurring serious delays) and (2) computer experimentation on developing agricultural systems, where experimentation involving the livelihoods of real people would not be possible.

THE REPAIR OF SYSTEMS

Although there is no advantage to a system operator in having greater knowledge than is required to operate it, this is only true while the system behaves in an orderly fashion. If it breaks down in some way, such as suffering from an outbreak of disease, then it is sensible to call in someone else, a veterinarian or a plant pathologist, for example, with the appropriate degree of knowledge and understanding.

If the breakdown is minor, it may be more economic if the operator is able to deal with it unaided; in many cases, therefore, it is an advantage for an operator to be using a model that is slightly more detailed than is strictly necessary just for smooth operation.

The knowledge required for substantial repair, however, may be much greater, and it is not possible to predict in advance exactly in what area this additional knowledge is going to be needed. That is why "repair" is commonly a separate professional function. But it is not enough for those primarily concerned with repair to concentrate on the component actually affected, even though this is the area in which they have been professionally trained.

Neither the problem nor the solution can be correctly framed with reference to one component part of the whole system. A "solution" to the problem has to result in an improvement to the whole system or the restoration of its former level of performance.

One of the most important systems principles is that the role and importance of a component can only be understood in relation to the system of which it is a part (Spedding, 1975). Thus, in an agricultural system, if one component fails, the choice of solution can only be made with reference to the whole system and its objectives. For example, if a cow suffers from a disease, there is no point in curing it in such a way that other cows are badly affected or that the pasture is destroyed or at such a cost that the farmer goes bankrupt. The same applies to preventive measures.

Now the "repair" expert may only know about the range of possible solutions and be quite unable to decide on the appropriate one. Ideally, in order to make his specialist contribution to whole-system functioning, he should be able to comprehend a picture of that whole system so that he can evaluate the consequences to it of any change or action that he may recommend. He also needs, therefore, a model of the whole system but, in this case, not to operate it but to see the implications of changes within his special competence to advise upon. Unfortunately, almost any such change may have implications for management, costs, labor, and a host of other important factors.

Sometimes, then, the specialist "repair" expert has to join forces with the operator in order to decide what should be done; this is also true of other specialists, including those involved in research aimed at system improvement.

THE IMPROVEMENT OF SYSTEMS

Sometimes the improvement of systems can be seen as an evolutionary progression. This ought to be so, very often, where subsistence farmers are being helped not merely to become better subsistence farmers (permanently) but gradually to evolve into commercial farmers as their output begins to exceed their own needs. In this sense there is a parallel with the evolutionary progression of some natural ecosystems.

In general, however, improvement means changing the system in ways that result in the more efficient use of resources for the achievement of the system objectives. Again, this frequently involves changes in one or a few components only, and the innovations are devised by specialists in those areas. The same problem is then encountered as to how to ensure that any change or innovation will actually improve the performance of the whole system. Furthermore, this does not simply apply to outputs. No one in agriculture wishes to increase output, except per unit of some resource (or several). Thus, an improvement may involve greater output, provided that inputs are not disproportionately increased, or a reduction in inputs, provided that outputs are not disproportionately decreased. It is generally the ratios relating outputs and inputs that have to be improved, thus constituting an increase in the efficiency of the system in the terms in which the operator has chosen to measure it.

The essential basis of any assessment of improvement has to be either a model (and this would include, e.g., a small experimental farm unit on which innovations can be tested) or a process of trial and error in the real world. The disadvantages and costs of the latter are obvious, but it is also the case that such experiments are uncontrolled and extremely difficult to evaluate. The results tend to be confounded by the multiplicity of factors influencing them, including the weather, and the interpretation of the findings is fraught with difficulty. Indeed, it is these kinds of difficulty that give rise to the need for research institutions because of the problems of farmers learning from their own experience.

The advantages of a model in this context are clear, but they depend upon the validity of the model, and it has to be recognized that model building and testing may be very costly. This is partly due to the fact that individual farms are so different that a large number of models is required. At this point it may be helpful to deal separately with these two issues: the validation of models and variation between farms.

THE VALIDATION OF MODELS

Validation simply means testing the validity of a model for a particular purpose; it can therefore only be tested in this context. The question posed is whether the model gives the same result as the real-world system operating under a relevant range of values for the main variables. It is very important to recognize in advance what is going to be meant by "the same." Real-world systems vary in their outputs, according to circumstances, so there is no one "correct" answer. Some kind of statistical variation has to be regarded as the "real" situation, and some range of variation about (or within) this has to be regarded as acceptable in a valid model.

All this relates, or course, to the internal workings of the model (the detailed testing of component processes and their quantification is more properly regarded as verification, however), and yet the output of an agricultural system may be just as affected by external variables. For example, some measure of profit will be an appropriate measure of the output of many agricultural systems but profit is influenced by externally determined costs and prices as well as by internal biological and technical processes. Frequently, the biggest variations are due to cost and price changes, and models of the systems themselves can only predict the consequences of such changes, not their probability.

Thus, a valid model is relevant for all such circumstances but cannot always help in choosing the most advantageous combination of controllable variables. In a sense cost and price changes are no different in principle from unpredictable changes in the weather, which affect biological and technical processes. Indeed, very often bad weather leads to good prices because total output is low, and good weather leads to low prices, a compensatory mechanism which looks different to the producer and the consumer.

In addition to all these variables, individual farms (and farmers) are very different from each other.

VARIATION BETWEEN FARMS

Where agricultural systems are weather dependent and greatly influenced by men, topography, and climate, each farm may be unique in important features. On the other hand, in intensive poultry enterprises the environment is closely controlled, the nature of the land is largely irrelevant (although this may affect waste disposal), and the degree of automation may greatly reduce dependence upon human skill. In such cases, individual variation may be minor and one model may have wide application, relatively independent of both site and even scale of operation. Where individual variation is considerable, however, and particularly where the husbandry or management skills of labor or managers has an important influence, models with general application may be of a rather limited, skeletal kind.

An interesting parallel exists between farms and organisms. Animals, for example, exhibit individual variation in response to factors such as climate, but they have evolved into groups that can be classified. Primitive agricultural systems have similarly evolved over long periods of time, and unsuccessful forms have failed to survive. But commercial agricultural systems have to change in response to an economic climate, which itself can change very rapidly and in quite extreme ways. Furthermore, whereas an animal greatly resembles its parents, so that change is relatively slow in both climate and animal, farms can change radically from their predecessors, in nature and scale, and can exist in a wide variety of intermediate forms. There are not just cattle farms and sheep farms, for example, but, in addition, mixtures in almost all conceivable proportions. The classification of agricultural systems is essential; otherwise it is not possible to say to what other (similar) systems the results of any one study relate, but the comparison can never be as distinct as in the case of animals and plants. There will thus be, except in the more standardized intensive systems with considerable environmental control, a great deal of variation among individual farms. This may often be no greater than the variation among years on the same farm, however, and sometimes this variation will be of the same kind. Examples are variations in yield due to weather differences among years on the same farm and due to topographical or managerial differences among farms. The same model may be able to cover such variations. But in the case of differences in the structure of systems, such as between farms with sheep and pigs and those with cattle and sheep, the very structure of the model will be different and contain quite different relationships. This contrast makes model construction more difficult but does not alter the fact that models are required.

MODELING IN AGRICULTURAL AND NATURAL ECOSYSTEMS

Both agriculture and ecology require models for the same basic reason: any study aims to build up a picture of the subject studied, of how processes operate, and of how changes in driving variables influence system output or behavior. In addition, agricultural models are the basis for activity, management, and intervention and help to decide on actions, to guide management, and to predict the consequences of intervention.

At first sight it would seem that the main difference between models of the two systems relates to the fact that every agricultural system is purposive, though not usually only concerned with one purpose, whereas natural ecological systems are not. However, every system can be viewed from a great many points of view, and in constructing a model a choice has to be made. There is no possibility of any model or picture actually representing all possible views of a system simultaneously (Spedding, 1979). The choice is usually clear for agricultural systems: the model has to reflect the purposes for which the system is studied. In the case of "operation," the purpose of the model, its structure, its outputs, and its inputs are related to the purpose for which the system is operated. If it is concerned with profit, then the model must contain those relationships that affect profit and omit those that are irrelevant or trivial in their influence.

Models of natural systems similarly have to reflect a point of view adopted by the observer. The observer may be interested in the flow of energy or carbon or water in a system, and a model can then be satisfactorily expressed in one or other of those terms. To express it in all terms simultaneously is not possible, but on the other hand expression does not have to be confined to one mode only. In models of agricultural systems, for example, money flows may usefully be combined with physical flows to describe complete chains of events.

Whatever may be regarded as the purpose of the system, therefore, or even whether it is judged to have one, a model of it must have a clear purpose; this will reflect the point of view from which it is being described. The degree of detail in which the model is constructed must, in turn, reflect this purpose.

The main difference between agricultural and ecological systems lies in the necessary emphasis in agriculture on outputs from the system. Natural ecosystems are largely self-contained for all but energy. Only primitive agricultural systems are of this kind, and, as the removal of outputs increases, there has to be a proportionate increase in inputs to match what is being removed. This feature of modern agriculture is primarily a consequence of the separation of the majority of the people from the agricultural systems that produce their food—in some cases by thousands of miles.

REFERENCES

Brockington, N. R. (1979). *Computer Modelling in Agriculture*. Oxford Science Publishers, Clarendon Press.

Spedding, C.R.W. (1975). *The Biology of Agricultural Systems*. Academic Press, London.

Spedding, C.R.W. (1979). *An Introduction to Agricultural Systems*. Applied Science Publishers, Essex.

The Linkage of Inputs to Outputs in Agroecosystems

George W. Cox

Department of Biology
San Diego State University
San Diego, California

INTRODUCTION

The essence of agriculture is use of energy and materials inputs to stimulate productivity of agroecosystems and concentrate it into forms useful to man. Productive agriculture, regardless of its nature, requires the use of concentrated resources to this end. Shifting cultivation, one of the most basic forms of nonmechanized farming, relies upon the exploitation of pools of nutrients and the ecological organization that have accumulated over a period of ecological succession. In pastoralism, herds of animals are moved from place to place, enabling them to exploit concentrated forage and water sources. In permanent farming and ranching, energy and materials resources are brought to the production site. Mechanized farming intensifies this last strategy and seeks economies of scale that enable the levels of such useful inputs to be increased to the most profitable point.

Intensive, mechanized agroecosystems differ from natural ecosystems in several quite significant ways, however (Fig. 1). The agroecosystem is less diverse in its species composition, and the community of higher plants and animals does not exploit niche space as fully as does the community in a

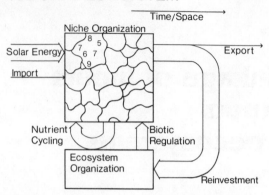

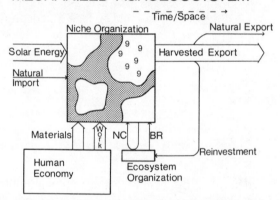

Figure 1. Natural ecosystems differ from mechanized agroecosystems in several ways. The biotic community of the natural ecosystem is more diverse (indicated by the number of cells in the niche space box) than that of the agroecosystem and exploits more fully the available niche space. The characteristics of individuals (genetics, age, health) within a species (indicated by numbers within one species' cell) tend to be varied in natural ecosystems, but nearly uniform in agroecosystems. Natural ecosystems are more continuous in space and time, and they reinvest the bulk of their production in their own ecosystem organization. The export of food from agroecosystems limits such reinvestment, and makes these systems dependent on materials inputs and work from the human economy.

natural ecosystem. Diversity within each species is low in the agroecosystem; individuals tend to be identical in their genetic composition and often nearly identical in size, age, and nutritional state. Whereas a mature natural ecosystem typically reinvests a major fraction of its productivity in maintenance of what we may call ecological organization, such reinvestment is very small in intensive agroecosystems because of the harvest and export of food and fiber. Ecological organization refers to the physical and biotic structure that functions to maintain high fertility and biotic stability in ecosystems. The job of providing the organization required for high productivity in an agroecosystem thus falls on man, and this job entails the use of concentrated inputs of energy and materials. In many agroecosystems, due to their discontinuity in space and time, large initial inputs are required to structure the productive system—planting the crop, orchard, or pasture. Productive inputs such as human and animal labor, machinery and the fuel it consumes, and fertilizers and other chemicals are thus essential to modern agriculture.

In his conscious and unconscious breeding of crops and domestic animals, man has enhanced the ability of these domesticates to convert such inputs into useful products. Primarily, however, these breeding efforts have affected internal allocation processes of domesticated species, enabling them to concentrate more of their assimilated food into desired structures or products (Evans, 1980). Rather than being r-selected or K-selected, these agricultural domesticates may be described as being A-selected (for allocation-selected). The allocation of productive potential to food organs or products, not surprisingly, is at the expense of many other characteristics that are adaptive in natural ecosystems. Tall stems, adaptive in competition with other plants, for example, are replaced by short stems, enabling the crop plant to allocate more assimilate to grain or tuber. Thus, in creating A-selected species, man assumes the responsibility for the ecological functions that have been sacrificed.

Productive inputs fall into three major categories in consequence: (1) those that replace quantities of essential nutrients and organic matter that have been harvested and exported; (2) those that serve to maintain essential physical features of the agroecosystem that would otherwise deteriorate because of reduced ecosystem reinvestment (tilth, hydrogen ion concentration, and salinity level, for example); and (3) those that replace the natural protection lost by A-selected species.

A GRAPHICAL MODEL OF INPUT ECOLOGY AND ECONOMICS

The total productivity of an agroecosystem can be related to the value (monetary, energetic, or other) of inputs as shown in Figure 2. Here, the value of output per unit of input is considered to show a unimodal curve with respect to total input level. This is equivalent to the postulate that total production is related to total input in a somewhat S-shaped fashion, with the greatest gain per unit input being in the middle range of total input, and with total produc-

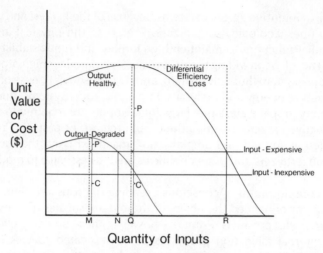

Figure 2. Output curves for healthy and degraded agroecosystems, and cost lines for expensive and inexpensive inputs. The symbol *P* indicates the profit (output value minus input cost) associated with a given input unit, the cost of which is designated *C*. See text for a full discussion of other relationships.

tion reaching an asymptote at some "ecological maximum." For simplicity, the cost of input units is displayed as a straight line at a certain height on the ordinate axis (some change in the unit cost of inputs might realistically occur with different total input levels).

Output curves for both healthy and degraded agroecosystems are shown in Figure 2. The degraded system possesses a lower maximum productivity and a lower maximum output per unit input. Furthermore, at the points of maximum difference between the output curves of healthy and degraded systems and the line for expensive input cost, the ratio of profit (*P*) to cost (*C*) is greater for the healthy system. In other words, this model predicts that in a healthy agroecosystem the maximum efficiency and the maximum profitability of inputs are greater than those in a degraded system.

The model shows that above some level of total production (*Q* on the output curve for a healthy system and the line for expensive inputs) the incremental effectiveness of inputs declines, eventually to the point (*R* on the input scale) at which the cost of additional inputs equals the value of added production. The latter is the point of maximum profitability of the agricultural enterprise. Additional production may still be achieved but only with input costs that exceed the value of the extra output. Eventually, as Odum et al. (1979) note, high levels of even usable inputs begin to stress the system, and incremental output can become negative.

Even within the profitability range, however, losses of productive inputs occur. A portion of these losses may be considered differential efficiency losses—the difference between the output curve and the maximum level of output per unit input. As Spedding (1975) has noted, the absolute size of differential efficiency losses increases with the total output of the system and may be linked to this output in such a way that attempts to reduce these losses may reduce total output as well.

Inexpensive (or subsidized) inputs permit production to be increased, so that the intersection of the output curve and cost line is moved closer to the ecological production maximum. The position of the cost line thus determines both the profitability of a given operation and the intensity of production that is economic.

This model suggests three basic strategies for increasing the net value of agroecosystem output: (1) changing the scale or shape of the output curve, as by crop breeding, choice of more productive crop types, or development of overyielding crop mixtures; (2) developing inputs that are more effective per unit cost, such as cheaper fuels or fertilizer formulations that permit greater recovery of nutrients by plants; and (3) improving the physical and biotic condition of the agroecosystem, as by protection against soil erosion, improvement of soil tilth, or the enhancement of biological control.

TRENDS IN INPUT EFFECTIVENESS

This model provides a guideline for discussion of trends in the effectiveness of materials and energy inputs to mechanized farming in the United States. Traditionally, efficiency in agriculture has been viewed in relation to labor costs. Now, however, it is becoming necessary to examine production in relation to other types of productive inputs (Evans, 1980), and we shall focus our attention on these.

Since the early years of this century, the value of farm production relative to purchased materials and energy inputs has declined steadily (Table 1). The only exceptions to this trend have occurred during wartime periods when the monetary value of production has risen sharply relative to the costs of inputs. The value of production has declined from about 4 times the cost of materials and energy inputs in the early 1900s to about 1.5 times these costs today. This decline in profit ratio has been offset largely by the increase in farm size, permitted by mechanization, that has enabled the absolute difference between production value and cost per farm to increase. To farm in a conventional fashion today, one *must* operate on a large scale to make an adequate total profit.

Table 1. Value of production per farm (U.S.) in relation to the costs of purchased materials, equipment, services, and repairs for periods from 1910 to 1979.[a]

Period or Year	Cash Value of Marketing and Home Consumption	Costs of Production Supplies, Equipment, Services and Repairs[b]	Production/ Cost	Production: Cost Increment Ratio
1910–1914	$ 7,147	$ 1,831	$ 3.90	$ —
1915–1919	12,471	3,005	4.15	4.53
1920–1924	11,691	3,329	3.51	−2.41
1925–1929	12,695	3,764	3.37	2.31
1930–1934	7,567	2,334	3.24	(3.59)
1940	9,592	4,204	2.28	1.08
1945	24,019	8,739	2.75	2.46
1950	30,524	14,603	2.09	6.19
1955	31,168	15,654	1.99	0.18
1960	35,453	19,309	1.84	1.17
1965	40,176	23,932	1.68	1.02
1970	51,290	31,444	1.63	1.48
1975	89,478	54,451	1.64	1.66
1979	132,906	87,079	1.53	1.33

[a] ESCS, 1979b.

[b] Current farm operating expense less hired labor, interest, dwelling repair; plus land and nondwelling capital improvements and the cost of trucks, tractors, and other equipment and machinery.

The effectiveness of increments of materials and energy in stimulating increased production has varied greatly for various periods since 1910 (Table 1). However, since 1960 the ratios of production gain to materials and energy increment have never shown the high values recorded at several earlier periods. Moreover, in the period 1975–79, when input costs increased more than in any other period, the increment ratio declined substantially from that for 1970–75.

The dependence of farm production on materials and energy inputs now varies significantly with farm size, as does efficiency, expressed as the value of marketed production relative to production costs (Table 2). With an increase in farm size there is a progressive decline in land value and in the absolute value of production per acre. In large measure (but not completely) this reflects the fact that large farms tend to be concentrated in areas where productivity per unit area is low—dry, cold, or mountainous regions. Increasingly, however, large farms are the result of consolidation of once-profitable small farms (Nuckton, 1980). The utilization of fertilizers, other chemicals, and direct energy supplies (fuels, electricity) per acre also declines with increasing farm size (except for a slight increase in per acre chemical usage on farms over 500 acres in size). However, when the value of these three types of inputs is calculated per dollar of marketed production, the importance of chemical and fertilizer inputs increases progressively with farm size; direct energy use, instead of decreasing, remains relatively constant with increased farm size. Other cost categories, including labor and customwork, exhibit more complicated patterns of variation but generally show their lowest cost for farms of intermediate size.

Altogether, the ratio of marketed production to total production costs (including labor) increases with farm size to a maximum for farms of 220–499 acres and declines for still larger farms (Table 2). The decline for these larger farms is due to increases in several cost categories, especially fertilizers, chemicals, labor, and customwork. While only 15% of U.S. farms fall in these larger categories, they contain 71.3% of total farm acreage. The differential efficiency with which such large farms use productive materials and energy inputs is thus of great consequence.

In terms of the graphical model presented earlier (Fig. 2), the intensity of utilization of materials and energy inputs in large-scale, modern farming has increased well beyond the point at which their incremental effectiveness is highest. Continued increase is now being promoted only by the smaller and smaller profit increments that can still be gained as production approaches the point of input–output equality.

For one of the most important productive inputs, fertilizer, this trend is particularly important. For the period 1910–49 the slope of the regression relating total U.S. fertilizer use to total crop production equalled 0.625; crop

Table 2. Marketed farm production value and production costs for 1978 in relation to farm size.[a]

	Farm Size (acres)								All Farms
	1–9	10–49	50–99	100–219	220–499	500–999	1000–1999	>2000	
Percent of farms	8.6	19.1	15.6	22.3	18.6	8.6	3.9	2.5	100
Percent of acres	0.1	1.3	2.9	8.6	15.7	15.2	13.7	42.4	100
Value per acre ($)	14208	2509	1276	1015	981	848	629	276	656
Marketed products per acre ($)	3774	489	219	177	166	140	104	42	111
Production costs per acre ($)	2424	321	133	101	88	75	56	24	62
Marketing/cost	1.56	1.52	1.64	1.75	1.90	1.86	1.85	1.74	1.79
Fertilizer $ per harvested acre	69	27	23	21	21	21	18	17	20
Chemicals $ per harvested acre	41	14	10	8	8	9	9	10	9
Direct energy $ per farm acre	220	27	12	10	9	8	6	2	6
Fertilizer $ per marketed $	0.007	0.021	0.039	0.051	0.064	0.075	0.076	0.055	0.059
Chemicals $ per marketed $	0.004	0.011	0.017	0.019	0.025	0.032	0.037	0.034	0.027
Direct energy $ per marketed $	0.058	0.055	0.057	0.057	0.056	0.058	0.061	0.052	0.056
Labor $ per marketed $	0.080	0.089	0.074	0.056	0.052	0.062	0.080	0.109	0.071
Customwork $ per marketed $	0.008	0.013	0.018	0.018	0.015	0.015	0.018	0.020	0.016

[a] Data from the 1978 U.S. Census of Agriculture.

production increased at a rate faster than fertilizer use (Fig. 3). For the period 1950–78, however, the slope of this regression has been 2.456, a highly significant increase ($t = 10.23$, DF $= 61$, $P < 0.001$). This means that almost four times as much fertilizer is now required to achieve the same increase in crop yield as in the early years of the century.

The effectiveness of fertilizer inputs is closely related to the condition of the soil, particularly the degree to which the topsoil has been depleted by erosion. In the mid-1970s soil erosion on U.S. cropland was estimated to average about 12 tons acre^{-1}, about three quarters of which was water erosion and one quarter wind erosion (Pimentel et al., 1976). The reductions of crop production in specific field situations, due to erosion impacts, are well known, and Pimentel et al. (1976) have estimated that 10–15% of the potential productivity of U.S. cropland has been lost because of erosion. Brink et al. (1977) have also shown that the erosion problem is an intensifying one, due to the recent expansion of cultivation onto marginal land and the shift toward continuous cultivation of certain high-value crops that also have a high erosion risk.

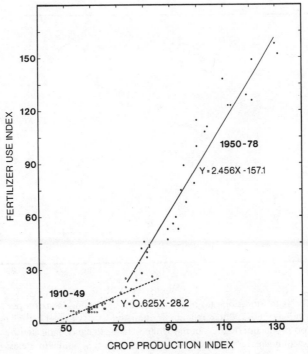

Figure 3. Regression relationships between total fertilizer use indices (*Y*) and total crop production indices (*X*) for the periods 1910–49 and 1950–78. Data from ESCS (1979a).

In the United States regional erosion levels are strongly correlated with the effectiveness of fertilizer inputs. In Figure 4 estimated soil erosion rates for USDA crop production regions for 1975 (derived from estimates for Water Resources Aggregated Subareas from SCS, 1977) are compared to the regression slopes of crop production per acre (index, 1967 = 100) on nitrogen fertilizer use per acre (index) for the period 1960–78. Soil erosion (sheet and rill) levels varied from a low of 1.11 tons acre^{-1} for the Pacific Region to 20.90 tons acre^{-1} for the Delta Region. Fertilizer effectiveness declines precipitously with increased erosion level ($r = -0.79$, $t = 3.65$, DF = 8, $P < 0.01$). In both the Delta and Appalachian Regions, per acre crop production showed an inverse relation to fertilizer use, indicating that even the considerable increases in fertilizer use per acre (range of 20.0–32.6 tons acre^{-1} in the Delta Region, 5.4–19.4 tons acre^{-1} in the Appalachian Region) that have occurred in these areas have been inadequate even to maintain a stable level of production.

Regression relationships for nitrogen fertilizer use and crop production can be used to estimate the reduction in U.S. crop production in 1978 due to

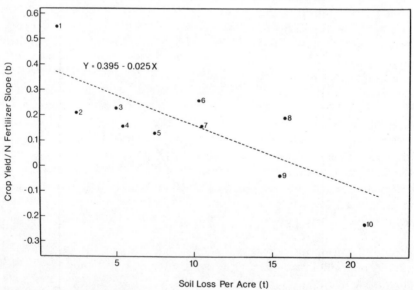

Figure 4. The regression slope (*b*) of per acre crop production index on per acre nitrogen fertilizer use index (1960–1978) in relation to estimated levels of sheet and rill erosion (1975) for USDA crop production regions of the United States. Regions, indicated by number, are: 1, Pacific; 2, Mountain; 3, Northern Plains; 4, Lake States; 5, Southern Plains; 6, Corn Belt; 7, Northeast; 8, Southeast; 9, Appalachian; 10, Delta. The regression relationship between yield–fertilizer slope (*Y*) and soil loss (*X*) is highly significant (*F* = 13.49, *P* < 0.01).

erosion effects. If erosion levels were everywhere as low as for the Pacific Region, estimates of regional crop production indices for 1978 can be calculated from the regional regression intercept, the 1978 regional fertilizer use index, and the slope (b) of the fertilizer-production regression for the Pacific Region. The ratio of expected to actual crop production indices can then be applied to regional production data for 1978 to estimate production at Pacific Region erosion levels (actual production data from ESCS 1979a). For the contiguous 48 states this analysis suggests that crop production was reduced from a potential $69.4 billion to $51.7 billion, a decrease of 25.4%. This estimate is considerably larger than that noted earlier and corresponds to a loss of about $53.34 per harvested acre of cropland.

Unfortunately, the short-term economics of farm production do not often provide strong incentive to implement soil conservation practices. Rosenberry et al. (1980), for example, found that even though existing rates of soil erosion in the Southern Iowa Conservation District were enough to reduce yields by an estimated $11.4 million annually, the implementation of various erosion control alternatives, all of which combined crop rotations with residue management, contouring, and/or terracing, would require the farmers of the region to forego income in amounts three to eight times as great. This essentially resulted from the substitution of low-erosion legume and hay crops for high-value crops with high erosion risk. Mitchell et al. (1980) analyzed the costs and benefits over a 20-year period for terrace construction in Illinois. They also concluded that farmers would sacrifice income by investment in terrace construction in all cases except on intensively managed, steep slopes with poor subsoil conditions. In essence, these analyses suggest that the costs of lost production, or of extra fertilizer needed to prevent such losses, are less than the costs of soil protection.

Neither of these studies considered the possible impacts of increased future costs of materials and energy inputs, nor the indirect and external costs associated with erosion. The analyses reflect the fact that the structure of U.S. agriculture is determined by the immediate relative costs of alternative inputs, which now still favor reduction in labor intensiveness (Edens and Koenig, 1980). The unfortunate consequence of this fact is the progressive degradation of agricultural land. At some point, this degradation will become of enormous economic significance because of the increased amounts of costly inputs needed to offset reduced fertility and restore degraded soil.

The efficiency of productive inputs is reduced in similar fashion by deterioration of the air environment. Major reductions in crop yields—far greater than those due to visual damage of harvested materials—are now known to result from ambient levels of oxidants (mainly ozone) and sulfur dioxide. The EPA-funded Ohio River Basin Energy Study led to the conclusion that in

1976 the most probable reduction in yields of corn, soybeans, and wheat totaled 258 million bushels for oxidant effects and more than an additional 3 million bushels for sulfur dioxide effects (Loucks and Armentano, 1982). For these three crops alone, projected losses to producers in the Ohio Basin for the period from 1976 to 2000 are between 10.3 and 12.3% of potential yields, a quantity valued at from $7.0 to $8.4 billion, depending on specific assumptions about energy development patterns (Page et al., 1982). In California the economic effect on consumers and producers of damage to eight major fruit and vegetable crops in the South Coast Air Basin in 1975 was estimated to be $103 million (Leung et al., 1982). Nationwide, 1980 crop losses due to air pollution were estimated to be about $1.8 billion (Jacobson, 1982). A large share of these losses in effect represents the cost of productive inputs that failed to realize their biological potential.

An additional factor relevant to the efficiency of use of inputs such as fertilizers is the sensitivity of A-selected crop species to environmental deviations from the optimal conditions for which they were bred. The question of whether or not crop plants have become more sensitive to deviations of climate and other factors from the optimum has not received extensive analysis. Stanhill (1976) found no evidence of increased sensitivity for wheat in England over the past 750 years. During this time, yields increased from about 0.5 tonne ha^{-1} to 3.4 tonnes ha^{-1}, but the coefficient of variation for annual yields remained at about 7%. Luttrell and Gilbert (1976) determined the probabilities, in a given year, of wheat and oat yields in the United States falling 5% or more below the long-term trend line. For the period 1937–68 they found, in both cases, that the probability was lower than for the period 1870–1928, suggesting that yield variations have become less erratic.

Such may not be the case for all crops, however (Table 3). Since World War II, corn yields in the United States have increased dramatically. For the period 1961–80 the average yield per acre was 81.9 bushels, almost double the 42.1 bushels for the period 1945–60. However, for the United States as a whole, and for most of the important corn belt states, the variance in yield per acre has increased significantly, and this increase has been great enough to produce small to moderate increases in the coefficient of variation of yield per acre. In Nebraska and Missouri, where the coefficient of variation for the period 1945–60 was quite high, due largely to high variability of rainfall, increased irrigation of corn has evidently contributed to a reduction in yield variability.

For India, as well, where total grain production has grown at a rate of about 3% annually over the past three decades, intensification of production has been accompanied by an increase in instability of annual yields (Hazell, 1982). For the period 1967–1968 through 1977–1978, the coefficient of variation of total cereal production was 6%, up from 4% for the period

Table 3. Variability of corn yield per acre for the United States and for selected states for 1945–60 and 1961–80.[a]

Region	1945–1960				1961–1980				
	Bu/Acre	b[b]	S²res[c]	CV[d]	Bu/Acre	b	S²res	CV	F[e]
United States	42.1	1.402	11.74	8.14%	81.9	1.871	53.77	8.95%	4.58[g]
Illinois	57.5	1.318	33.41	10.05%	96.4	1.624	125.60	11.62%	3.76[f]
Indiana	55.5	1.029	19.26	7.91%	93.3	2.009	114.23	11.46%	5.93[g]
Wisconsin	51.8	1.450	22.58	9.17%	84.3	1.371	100.64	11.90%	4.46[g]
Nebraska	33.8	1.266	87.67	27.70%	81.0	2.516	131.31	14.15%	1.50
Kentucky	38.7	0.928	22.76	9.09%	74.8	1.847	110.53	14.05%	4.85[g]
Missouri	40.1	1.350	74.72	21.56%	70.7	0.913	197.34	19.87%	2.64

[a] Data from USDA annual *Agricultural Statistics* volumes.
[b] Slope of regression of yield per acre on year.
[c] Residual variance for regression relation of yield per acre on year.
[d] Coefficient of Variation of yield per acre.
[e] Ratio of residual variance for 1961–80 over that for 1945–60.
[f] Significant at $P = 0.05$.
[g] Significant at $P = 0.01$.

1954–1955 through 1964–1965. In regions such as this, interruptions in the availability of needed inputs such as fertilizers and power for irrigation pumping, as well as unfavorable weather and pest outbreaks, may contribute to yield instability.

These examples suggest that many modern high-yielding crop varieties are really much more dependent on near-optimum conditions than older varieties, and that the reductions in year-to-year variability of their yields are due largely to the more intensive use of elaborate, input-consuming practices to maintain optimum conditions. Were modern varieties cultivated under the practices of earlier decades, their performance would likely be much more erratic than that of older varieties.

INCREASING THE EFFECTIVENESS OF PRODUCTIVE INPUTS

The possible approaches to increasing the efficiency of materials and energy inputs relate to the three functions of agricultural input noted earlier. The first of these is extending the range and height of the output curve for a particular system (Fig. 2). Traditional plant and animal breeding programs that modify the allocation of assimilated energy to desired products will continue to be an important means of accomplishing this goal. However, the gains in improved allocation must be weighed against the increased need for protective inputs for these A-selected forms.

More basic approaches to improvement of the productive physiology of crop and domestic animal species must also be sought. One approach that offers interesting potential is the introduction of C_4 photosynthetic pathways (and the companion relationships of reduced photorespiration) into C_3 species. Although it is uncertain whether full transfer of such physiological systems can be achieved, significant reductions in the photorespiratory losses experienced by C_3 forms may be possible (Bassham, 1977). Innovative techniques of plant genetics, such as cell culture, protoplast fusion, and recombinant DNA, offer new opportunities for basic physiological modification of crop species (Day, 1977). For livestock species increased reproductive output of animals maintained for breeding purposes is an important goal (Wilson, 1973) since the maintenance costs of breeding stock are a major determinant of meat production efficiency.

A different approach to the improvement of output curves for agroecosystems is the development of overyielding crop and animal combinations (Kass, 1978; Lin, 1982). Mixed cropping has only recently received serious study, but it is apparent that significant yield advantages often exist, especially when there are pronounced differences among species in growth form and/or maturation date (Baker, 1979; Fisher, 1979). Interest in mixed cropping has been greatest in developing countries where farming has not yet become highly mechanized, but the basic approach can also be adapted to many mechanized agroecosystems. Strip cropping, alternate row cropping, and interseeding of grain and cover crops are being practiced with success by mechanized techniques. In China polycultures of species of carp and other fish have long been farmed in ponds, lakes, rivers, and rice paddies. Refined systems of multigrade polyculture are now being developed; in these, fish are sorted to ponds by size, so that the total fish biomass can be adjusted closely to the productive capacity of the pond (Lin, 1982).

A second approach to increasing the efficiency of productive inputs is the development of ways to make inputs more effective per unit cost. New, more active, and less expensive chemicals of various types can certainly be produced. Much opportunity still exists, for example, to improve fertilizer formulations to achieve an optimal balance of nutrients, including micronutrients, and an optimal rate of release of nutrient elements into available form. However, major improvements can also be made in the utilization of the materials and technology now available. Although the trend toward reduction of labor in farming continues, the increasing costs of fuels and petroleum derived chemicals have stimulated the growth of consultant services dealing with farm operation, maintenance of soil fertility, and pest management. In the San Joaquin Valley of California, pest management consultants have enabled cotton and citrus growers to cut their insecticide use by over 50%, while gaining greater yields per acre (Hall et al., 1975). The creation of skilled labor in areas such as this is one of the quickest ways that efficiency in the use of productive inputs can be increased. Stimulating the growth of such professions by the provision of training programs, scholarships, and educational loans should be a priority for federal and state governments.

Finally, and most significantly from an ecological standpoint, input efficiency can be increased by strengthening ecosystem characteristics that control fertility and productivity. In the humid tropics this approach—the retention of the basic structure and process that operate in mature natural systems—is probably the only realistic means of developing crop ecosystems with sustainable productivity (Cooper, 1981). The capacity of tropical lowland soils to release nutrients by weathering is low, their ability to retain nutrients in available forms weak, and the risk of nutrient loss by leaching great. Neither are animal pests and weeds periodically reduced to the same extent that they are by winter weather in the temperate zone.

Almost everywhere, however, protection and improvement of the soil against erosion damage is of foremost importance. Conservation tillage systems have become an important means to this end (Unger and McCalla, 1981). In 1979 conservation tillage was being practiced on 26.4% of U.S. cropland (23.9% minimum tillage and 2.5% no-tillage). Reduced tillage exhibits a number of effects, but wind and water erosion are greatly reduced, often to a negligible level. When soil moisture is limiting, yields are often higher under reduced tillage, and when moisture is adequate, soils well drained, and fertilization and weed control good, they usually equal those of conventional tillage (Unger and McCalla, 1981). Corn yields per unit of nitrogen fertilizer are reported to be much greater under no-tillage than under conventional tillage (Phillips et al., 1980), although leaching and denitrification can cause high losses of nitrogen in wet or poorly drained soils. The energy and dollar costs of grain production by no-tillage are also significantly less, due mostly to reduced machinery and fuel needs. For U.S. corn and soybeans total energy input was estimated to be 7 and 18% lower, respectively, than for conventional tillage (Phillips et al., 1980). In England the costs of cropping of cereals by no-tillage were estimated to be about half those of conventional cropping (Unger and McCalla, 1981). Reduced tillage requires effective weed control and thus requires large inputs of herbicides.

The huge economic losses due to soil erosion justify greater governmental investment. The question, of course, is how to obtain the needed funds. Because of the growing importance of U.S. food exports, the suggestion for a tariff on such exports, earmarked for soil conservation, deserves consideration (Seitz, 1981). A 1% tariff would yield about $200–370 million annually. A soil management practice tax, with heavier taxes on acreage subject to tillage or cropping that favor erosion, may also be practical and effective (Heady and Daines, 1982).

Integrated pest management, with varied strategies of cultural and biological control (Batra, 1982), strengthens the regulatory capacity within agroecosystems. This often increases productivity, as well as reducing the need for pesticides. Adkisson et al. (1982) showed impressively how integrated control can improve both yield and net profit per acre for cotton in southern Texas. By planting cotton in narrowly spaced rows, forcing it to mature early by cultural practice, and severely limiting pesticide use (to avoid secondary pest outbreaks), yields were increased from 625 to 765 lb acre^{-1} and net profit from $109 to $252 acre^{-1}. Other practices were also included in the management system to reduce the abundance of overwintering pest insects.

There is strong evidence that investment in integrated control has a greater benefit–cost ratio than investment in conventional chemical control. Pimentel et al. (1980) concluded that conventional chemical control returned about $4 for each dollar invested if only the direct costs of pesticides were considered, or about $3 if indirect and external costs were also counted. DeBach (1972) argued that the return for classical biological control was much higher, perhaps 30 : 1.

A revival of interest in organic farming, now practiced on perhaps 20,000 U.S. farms (USDA Study Team on Organic Farming, 1980), has stimulated a number of studies of the productivity and efficiency of this approach. Organic farms in the midwestern United States depend heavily on crop rotations that include legumes to supply nitrogen for other crops in the cycle. This practice also tends to improve soil organic content and to provide protection against erosion, as well as reduce the input of costly nitrogen fertilizers (Lockeretz et al., 1981). The energy intensiveness of midwestern organic farms is considerably less than that of conventional farms, the energy consumption per dollar of marketed product ranging from 38 to 45% of that for conventional farms between 1974 and 1978 (Lockeretz et al., 1981). Although yields per acre of certain crops, particularly corn, are often somewhat less under organic practice, the lower costs per acre make organic farms competitive with conventional farms in net profits (Lockeretz et al., 1981). These observations show that organic techniques deserve serious study and indicate that certain organic practices may contribute both to the improvement of agroecosystem health and to direct increase in the efficiency of use of energy and materials inputs to farming.

The political and economic forces that influence the use of materials and energy inputs in U.S. agriculture are complex and powerful, however. These range from the policy objectives of executive branches of the federal government, concerned with matters such as the role of food exports in maintaining a favorable balance of trade to the lobbying activities of farm producer groups, agribusiness interests, and consumer organizations before congressional committees and state legislatures. Agribusiness, which has as its first interest the sale of machinery, fuels, chemicals, and other products, carries its message directly to the farmer through sophisticated advertising. The interests of these various groups are often in conflict, but the overall result of these forces has been to increase production by high-input technology to the point where the price received often does not cover the investment and production costs (Johnson and Quance, 1972).

AN ENERGY EFFICIENCY INDEX

Under given extrinsic controls of climate and substrate, natural ecosystems have developed characteristic levels of primary productivity. In converting these natural ecosystems to agroecosystems, man changes the level of primary production, usually (but not always) reducing it to a lower level. By careful design and wise use of available productive inputs, however, man should be able to create agroecosystems that function at the levels of productivity defined by ecological controls (or perhaps even better!).

Thus, we may suggest that intensification of agroecosystem productivity should be strongly encouraged up to the level shown by the natural system it replaced, and that this investment should also seek the greatest efficiency in the use of inputs rich in fossil fuel energy (as opposed to human and animal labor and energies from flow sources). From this argument we can derive an efficiency expression, the Energy Efficiency Index (EEI):

$$EEI = \frac{\text{agroecosystem } NPP}{[(FF + D)(T + D)]^{0.5}}$$

where *NPP* = net primary production;
       FF = fossil fuel input;
        T = total energy input;
       D = the excess of natural ecosystem *NPP* over agroecosystem *NPP* ($D \geq 0$).

This index increases in value as agroecosystem NPP rises toward that of the replaced natural system, as total energy input for a given agroecosystem NPP is reduced and as fossil fuel energies are reduced relative to other energy inputs. Above the NPP level of the natural ecosystem, only the latter two relationships carry weight (when *D* has dropped to zero). This expresses the philosophy that intensification to unnaturally high levels should be attempted only when the gains in production are considerable per unit of additional energy input.

Using this expression we can calculate EEIs for the various rice production systems considered by Freedman (1980). These systems are traditional, transitional, Green Revolution, and experimental (Table 4). The traditional system was in use in the precolonial period in southeast Asia. The transitional system integrated basic farm tools into production in a labor-intensive fashion. The Green Revolution system emphasizes the intensive use of fertilizers, insecticides, herbicides, and modern machinery. The experimental system embodies approaches that exemplify intermediate technology but are geared to achieving high productivity by high-yielding rice varieties.

Freedman (1980) notes that the energy efficiency of the experimental approach exceeds that of the Green Revolution system, and we see that the EEI for this system is the highest of the four approaches considered (Table 4). Thus, this index, or one like it, may provide a useful tool for evaluating the progress of agricultural technology toward the goal of creating productive systems that make efficient use of materials and energy inputs.

SUMMARY

In modern agriculture various materials and energy inputs are required to replace exported materials, to maintain productive conditions, and to pro-

Table 4. Energy efficiency indices (EEI) for rice production systems in an area of moist tropical forest climax vegetation. Data on yields and energy inputs from Freedman (1980). See text for characterization of the different production systems.

System	Crops per Year	Annual Energy Input (kcal m^{-2})		Agroecosystem Net Primary Production (kcal m^{-2})[a]	Deficit (D) (kcal m^{-2})[b]	EEI[c]
		Fossil Fuel (FF)	Total (T)			
Traditional	2	1.4	65.6	1815.0	6705	0.27
Transitional	2.5	142.0	366.0	4860.0	3660	1.24
Green Revolution	3	1626.3	1648.4	9583.2	0	5.85
Experimental	3	1012.8	1034.9	9583.2	0	9.36

[a] Assumes that total rice ecosystem NPP is two times grain production.

[b] Assumed NPP of moist tropical forest (8520 kcal m^{-2} yr^{-1}) minus agroecosystem NPP.

[c] $EEI = \dfrac{\text{Agroecosystem NPP}}{[(FF+D)\,(T+D)]^{0.5}}$

205

tect allocation-selected crop and livestock species. In U.S. agriculture the incremental effectiveness of such inputs has declined as farms have increased in size, adopted more mechanized techniques, and become more intensive in operation. The effectiveness of additional increments of fertilizers, in particular, has declined. In addition, the effectiveness of increased fertilizer use is significantly lower in regions with high levels of soil erosion and probably also in regions with high levels of oxidant and sulfur dioxide air pollution. Finally, both in the United States and elsewhere, there is evidence that some modern crop varieties, such as corn, are more vulnerable to environmental variation than older varieties, and thus more dependent on high levels of protective inputs.

The efficiency of materials and energy inputs can be improved by increasing the productive potential of agricultural species, by developing more effective inputs, and by improving agroecosystem health. Breeding efforts directed at basic aspects of the photosynthetic and respiratory physiology of crops and efforts to create overyielding crop and animal polycultures are two important approaches to increasing the productive potential of agroecosystems. Improved farm chemicals, together with increased use of skilled consultant services, can improve the efficiency of materials and energy inputs considerably. Efforts to develop farming practices that protect and improve soil conditions and that possess a high degree of intrinsic biological control of pests are sound approaches to improving the health of agroecosystems.

Evaluation of the efficiency of inputs should realistically consider the natural productivity patterns of native ecosystems. Attempts to raise agroecosystem productivity to levels far above those of native systems are justified only when this can be done with a high degree of efficiency.

ACKNOWLEDGMENTS

I thank Charles F. Cooper, David A. Farris, Garfield House, Richard Lowrance, Dwain W. Parrack, and Ben Stinner for comments and suggestions on earlier drafts of the manuscript.

REFERENCES

Adkisson, P. L., Niles, G. A., Walker, J. N., Bird, L. S., and Scott, H. B. (1982). Controlling cotton's insect pests: A new system. *Science* **216**:19–22.
Baker, E. F. J. (1979). Mixed cropping in northern Nigeria. III. Mixtures of cereals. *Exp. Agr.* **15**:41–48.
Bassham, J. A. (1977). Increasing crop production through more controlled photosynthesis. *Science* **197**:630–638.

Batra, S. W. T. (1982). Biological control in agroecosystems. *Science* **215:**134–139.

Brink, R. A., Densmore, J. W., and Hill, G. A. (1977). Soil deterioration and the growing world demand for food. *Science* **197:**625–630.

Cooper, C. F. (1981). Climatic variability and sustainability of crop yield in the moist tropics. In *Food-Climate Interactions*. W. Bach, J. Pankrath, and S. H. Schneider (eds.), D. Reidel Publishing Co., Dordrecht, Netherlands, pp. 167–186.

Day, P. R. (1977). Plant genetics: Increasing crop yield. *Science* **197:**1334–1339.

DeBach, P. (1972). The use of imported natural enemies in insect pest management ecology. *Proc. Tall Timbers Conf. on Ecol. Animal Control by Habitat Management*, No. 3, 211–233.

ESCS. (1979a). Farm income statistics. USDA, Economics, Statistics, and Cooperatives Service, Statistical Bull. No. 627.

ESCS. (1979b). Economic indicators of the farm sector: Income and balance sheet statistics, 1979. USDA, Economics, Statistics, and Cooperatives Service, Statistical Bulletin No. 650.

Edens, T. C., and Koenig, H. E. (1980). Agroecosystem management in a resource-limited world. *BioScience* **30:**697–701.

Evans, L. T. (1980). The natural history of crop yield. *Am. Sci.* **68:**388–397.

Fisher, N. M. (1979). Studies in mixed cropping. III. Further results with maize-bean mixtures. *Exp. Agr.* **15:**49–58.

Freedman, S. M. (1980). Modifications of traditional rice production practices in the developing world: An energy efficiency analysis. *Agro-Ecosystems* **6:**129–146.

Hall, D. C., Norgaard, R. B., and True, P. K. (1975). The performance of independent pest management consultants in San Joaquin cotton and citrus. *Calif. Agr.* **29:**12–14.

Hazell, P. B. R. (1982). Instability in Indian foodgrain production. *Int. Food Policy Res. Inst. Res. Report* 30.

Heady, E. O., and Daines, Jr., D. R. (1982). Short-term and long-term implications of soil loss control on U.S. agriculture. *J. Soil Water Conserv.* **37:**109–113.

Jacobson, J. S. (1982). Economics of biological assessment. Introduction and summary. *J. Air Pollution Control Assoc.* **32:**145–146.

Johnson, G. L., and Quance, C. L., eds. (1972). *The Overproduction Trap in U.S. Agriculture*. The Johns Hopkins University Press, Baltimore.

Kass, D. C. L. (1978). Polyculture cropping systems: Review and analysis. *Cornell Int. Agr. Bull.* **32:**1–69.

Leung, S. K., Reed, W., and Geng, S. (1982). Estimations of ozone damage to selected crops grown in southern California. *J. Air Pollution Control Assoc.* **32:**160–164.

Lin, H.-R. (1982). Polycultural system of freshwater fish in China. *Can. J. Fish. Aq. Sci.* **39:**143–150.

Lockeretz, W., Shearer, G., and Kohl, D. H. (1981). Organic farming in the Corn Belt. *Science* **211:**540–547.

Loucks, O. L., and Armentano, T. V. (1982). Estimating crop yield effects from ambient air pollutants in the Ohio River Valley. *J. Air Pollution Control Assoc.* **32:**146–150.

Luttrell, C. B., and Gilbert, R. A. (1976). Crop yields: Random, cyclical, or bunchy? *Am. J. Agric. Econ.* **58:**521–531.

Mitchell, J. K., Brach, J. C., and Swanson, E. R. (1980). Costs and benefits of terraces for erosion control. *J. Soil Water Conserv.* **35:**233–236.

Nuckton, C. F., ed. (1980). *Farm-Size Relationships, with an Emphasis on California.* Department of Agricultural Economics, University of California, Davis.

Odum, E. P., Finn, J. T., and Franz, E. H. (1979). Perturbation theory and the subsidy-stress gradient. *BioScience* **29:**349–352.

Page, W. P., Arbogast, G., Fabian, R. G., and Ciecka, J. (1982). Estimations of economic losses to the agricultural sector from airborne residuals in the Ohio River Basin region. *J. Air Pollution Control Assoc.* **32:**151–154.

Phillips, R. E., Blevins, R. L., Thomas, G. W., Frye, W. W., and Phillips, S. H. (1980). No-tillage agriculture. *Science* **208:**1108–1113.

Pimentel, D., Terhune, E. C., Dyson-Hudson, R., Rochereau, S., Samis, R., Smith, E. A., Denman, D., Reifschneider, D., and Shepard, M. (1976). Land degradation: Effects on food and energy resources. *Science* **194:**149–155.

Pimentel, D., Andow, D., Dyson-Hudson, R., Gallahan, D., Jacobson, S., Irish, M., Kroop, S., Moss, A., Schreiner, I., Shepard, M., Thompson, T., and Vincent, B. (1980). Environmental and social costs of pesticides: A preliminary assessment. *Oikos* **34:**126–140.

Rosenberry, P., Knutson, R., and Harmon, L. (1980). Predicting the effects of soil depletion from erosion. *J. Soil Water Conserv.* **35:**131–134.

SCS. (1977). Cropland erosion. USDA, Soil Conservation Service, Washington, D.C.

Seitz, W. D. (1981). A conservation tariff. *J. Soil Water Conserv.* **36:**120–121.

Spedding, C. R. W. (1975). *The Biology of Agricultural Systems.* Academic Press, London.

Stanhill, G. (1976). Trends and deviations in the yield of the English wheat crop during the last 750 years. *Agro-Ecosystems* **3:**1–10.

Unger, P. W., and McCalla, T. M. (1981). Conservation tillage systems. *Adv. Agron.* **33:**1–58.

USDA Study Team on Organic Farming. (1980). Report and recommendations on organic farming. USDA, Soil Conservation Service, Washington, D.C.

Wilson, P. N. (1973). Livestock physiology and nutrition. *Phil. Trans. Roy. Soc. London Ser. B* **267:**101–112.

Toward a Unifying Concept for an Ecological Agriculture

Wes Jackson

The Land Institute
Salina, Kansas

It is obvious that long-term food security for the world cannot be hitched to nonrenewable resources such as fossil fuels and mined minerals and that we will have to look elsewhere for the energy and materials to power civilization, including agriculture. We can only speculate as to the extent that the recent interest in an ecologically sound agriculture stems from a growing uneasiness about the vulnerability of the industrial world, including our industrial agriculture which requires high-quality fuels. Though industrial, or what I will call production agriculture is now the dominant means of food production in the developed countries; it is increasingly recognized as standing over and against sustainable, regenerative, or ecological agriculture. It is important, therefore, to examine differences. At one level, at least, their differences are more striking than their similarities.

The differences in points of view between production and sustainable agriculturists ultimately have to do with both time and faith. "Production" agriculturists either discount more of the future than "sustainable" agriculturists or have the faith that adjustments can be quickly made when the supply of fuel and materials runs low. In other words some production agriculturists believe we can quickly shift to solar energy or be saved by nuclear power, whether of the fission or fusion variety. Others do not care or are willing to let the future take care of itself.

Those who push for a sustainable agriculture, on the other hand, discount less of the future and, overall, are more conservative or skeptical that society can adjust quickly and safely to a changing future. When we explore other major features in a comparison of production to sustainable agriculture, we soon discover that few absolutes exist. For example, proponents of production agriculture have to be interested in some measure of sustainability even if it is just for one year. Plant breeders must have an eye to some level of sustainable yield as they select for resistance to various insects or pathogens. In this sense they are interested in more than production or yield; it is just that they have a shorter time frame than some of us would like, which gets us back to our earlier discussion. Sustainable agriculturists in turn must have some interest in production, for there is little purpose in farming if it does not yield something of human benefit. When we get too specific or insist on absolutes, distinctions fade.

It will be disastrous for both agriculture and culture if we allow these distinctions to fade because they are both real and crucial. I am one of those interested in an ecologically sound agriculture and, as such, want to talk about a conceptually different way of growing food. Those who share this value hope and expect that it is a way that we can run agriculture on sunlight, cut soil loss below or equal to biological replacement levels, and keep the soil healthy besides.

THE UNIFYING CONCEPT OF MODERN BIOLOGY

Before *On The Origin Of Species* was published in 1859, the world of the professional biologist stood before him like some large post office. The preserved and living plants and animals arriving in crates and boxes at the docks and depots of such places as London and Munich and Paris and Cambridge from around the globe were eventually sorted into what were regarded as the proper slots by the museum biologists of the time. If different enough, taxonomists assigned them a new name and a new slot. Species that differed but little were given slots adjacent to one another, much as a stamp or coin collector would order his collection. Individual representatives of thousands of species looked out on this world like unsigned portraits in a huge art gallery. It was an assembly with incomplete sense, for while there were thousands of clusters, the discontinuities between clusters were more striking than the similarities within them. In the schools devoted to medicine, professors and students of anatomy and physiology were no better off. They, too, noted similarities and differences but mostly left it at that. In one sense the living world stood as vertical strata—pigeon holes—in the human mind, and for the Western mind, at least, the major gap was between the human and the rest of creation, a gap created by the Creator's pause. For on the authority of Genesis, after the Creator had feasted His eyes on the creation before Him, He declared it good and created the human to be His foreman with explicit instructions to protect His interests.

And so, whether these creatures were scurrying crabs in a rocky tide pool or insects pinned inside a fumigated tray, green things gracing an arboretum or moth-balled in the herbarium, or ungulates running wild on an African savanna or mute and still in a glass case in London, they were understood first as representing individual species and secondarily as fitting into some ecological relationship. But that is as far as it went. Mostly they were regarded as autonomous units ordered on shelves or in trays and cabinets by the minds of men, more for convenience sake than for any larger understanding. Biologists of the Western world had more or less accepted the idea of instant creation. But as the naturalists went forth and returned, internal inconsistencies in this idea began to emerge. Large questions were asked and went unanswered: How, for instance, did all those marine fossils get buried in the strata high in the mountains if everything began around 4000 B.C.? As internal inconsistencies continued to pop up in the "instant creation" paradigm, the minds of biologists were being prepared for what was to become the era of modern biology.

The dyspeptic and shy Charles Darwin, drawing on his Galapagos experience, the essay of Parson Malthus, the geology of Lyell and Hutton, and his own voluminous notes, suddenly united all these life forms with the thread of time in his well-documented explanation of the workings of evolution through natural selection. Darwin tilted the vertical strata in the minds of cataloging biologists exactly 90°. With these strata resting comfortably now in a horizontal position, the time dimension was quickly added and, for the most part, gracefully internalized in the biologist's mind. Gaps that stood as discontinuities representing God's attention to specifics in creation when vertical, became understandable, though often bothersome, breaks in the horizontally laid fossil record. The new horizontal mind of biologists quickly fell into synchrony with the ancient strata of geology.

Not only did Darwin's unifying concept for biology bring more complete sense to the world of the cataloger, but the implications of this worldview also leaked over into physiology and anatomy and all the other disciplines of biology. No longer would the living world be simply regarded as consisting of individuals or species or even as part of an ecological setting. The creatures of the living world were more than that. They were products of forces that had shaped them over time. Though there were speculations on the origin of species before Darwin, his exhaustive treatment of the subject became the watershed. The dawn of modern biology arose in a London bookstore on a November morning in 1859.

THE NEED FOR AN INTERNALLY CONSISTENT
UNIFYING CONCEPT FOR AGRICULTURE

Darwin may have changed the minds of "pure" scientists in biology, but if he changed the minds of applied biologists working in agriculture, it was not reflected in their work. The approach of agricultural researchers to crop improvement scarcely budged until after 1900, after the rediscovery and expansion of Gregor Mendel's laws of heredity, nearly a half-century after *On The Origin Of Species* was published. It was Mendel's work that caught their fancy. They were more interested in the manipulations of heredity for crop improvement than they were in the ecological implications of Darwin's ideas. By paying more attention to Mendel's contribution rather than to the implications of the contributions of both Darwin and Mendel, they continued to operate as though agriculture could be understood in its own terms. The rules of heredity simply gave them the necessary knowledge to better manipulate domestic plant and animal populations—our crops and livestock, not ecosystems.

It is easy to appreciate how these researchers believed that agriculture could be understood in its own terms. After 10,000 years and more, since the beginning of agriculture, crops and livestock had become so streamlined for special human purposes through breeding that their roots in a former ecological setting were mostly irrelevant. The demands placed on these creatures by human ecology had taken precedence over the demands of natural ecology for so long that crops were regarded more as the property of humans than as relatives of wild creatures.

But the problem runs even deeper. Like many species, we have rather specialized food requirements. Unlike any other species, however, we have the ability to radically alter the environment in order to meet those requirements; and for that reason, primarily, we have become a problem for the earth and all of its life forms. We really were not much of a problem for the earth until we began to substitute single-species populations (crops) for diverse ecosystems. An area planted to plant populations met our specialized demands better than the generalized ecosystems of nature, especially in temperate regions.

We seldom appreciate how narrow our food requirements really are. But of the 350,000 plant species worldwide, only two dozen are of particular importance to us for food. Of the top 18 sources, 14 come from but two flowering plant families, the grasses and legumes (Table 1). Nine of the top 14 are grasses, and in all but one case (sugar cane) we are interested exclusively in the seeds. Furthermore, *Homo sapiens*, who would more appropriately be named Grass–seed eater, also relies heavily on land vertebrates for meat, which in turn depend heavily on grass as forage. It is no mystery that

Table 1. Twenty plant species (out of 350,000 worldwide). Grouped by family, ranked by importance.[a]

Poaceae (grasses)	Fabaceae (legumes)	Solanaceae	
1. Rice	8. Soybeans	4. Potatoes	
2. Wheat	15. Peanuts		
3. Corn	16. Field beans	Convolvulaceae	Musaceae
5. Barley	17. Chick peas	6. Sweet potatoes	19. Bananas
9. Oats	18. Pigeon peas		
10. Sorghum		Euphorbiaceae	Palmaceae
11. Millet		7. Cassava	20. Coconuts
12. Sugar Cane			
14. Rye		Chenopodiaceae	
		13. Sugar beets	

[a] From Wittwer, 1981.

the prairie states, the grassy states, from Ohio to the Rockies, from Canada to Texas, supply most of the food for the United States and ship most of the tonnage of a huge export market each year. More than any other family of plants, it is grass that supports us, for when we get serious about food production for meeting basic human needs, most of our field acreage features grasses. Isaiah exaggerated only slightly when he said, "All flesh is grass."

Though polycultures have been important to many indigenous peoples in the tropics, in the Orient, and among American Indians who grew corn–squash–bean associations, over most of the landscape we have featured monocultures. This is because, in most cases, we can understand and manipulate populations better than diverse ecosystems.

Nearly all the crops listed in Table 1, especially those which are the most important, are grown in monoculture. Some have been grown in highly simplified polycultures such as corn–bean–squash associations. Overall, the most extreme thinking beyond the population level has been to think of these crops as representative of a particular plant family, not as members of an ecosystem. But if we are to be serious about an ecologically sound agriculture, we have to think about plants as part of an ecosystem and as relatives of wild species who still have their ecological genes intact. When we think of the vegetative structure of our major crops and then think of wild ecosystems, two principle domestic ecosystem types feed humans—highly simplified prairies and highly simplified marshes. The vegetative structure of most of our farmed acreage, even in the wooded areas of our country, more nearly resembles the prairie than any other wild ecosystem. In the corn belt a corn field is a "tall grass" monoculture we have substituted for the grasses of the tallgrass prairie, which grew there before the widespread planting of

corn. A wheat field is a "midgrass" monoculture substituted for midgrass prairie, but it requires irrigation or summer fallow.
stituted for midgrass prairie, but it requires irrigation or summer fallow.

Marsh ecosystems are places where the biomass turnover is high (Odum, 1971). It is no wonder then that rice, a high-yielding grass grown in a "domestic marsh," is the primary crop of humans in the most crowded parts of the world. The problem is that if nature were fully in charge where rice and wheat are grown, the biotic diversity would be staggering. We homogenize the environment in order to feed from extremely narrow domestic prairies and marshes. We do not live by either wheat or rice alone, but we come uncomfortably close, and we could very well elect to homogenize the earth's habitat even more than it now is to meet the needs and demands of a growing global population.

It may be all right to have food demands that rely most heavily on grasses and legumes, but we had better take care to retain the variety within those two plant families to better enhance the long-term stability of agroecosystems. There is enough generic and species diversity within those two families alone to have domestic marshes and prairies over the landscape providing most of the necessary ingredients for our diet. We are fortunate that prairies feature grasses. The native prairie plants of central Kansas are 95% grasses; half of the remaining 5% are legumes (Weaver, 1954). A Wisconsin prairie may be only 60% grass, but grass still predominates (Ahrenhoerster and Wilson, 1981). The legumes, although they do not dominate any major ecosystem, have an abundance of generic and species diversity to draw on. There is no reason, therefore, to assume that we cannot begin to think of developing a sustainable agriculture at the ecosystem level rather than at the population or crop level in order to meet both the needs of the land and the needs of the people. We are fortunate in this, for what if no ecosystem existed in which at least one of our most important plant families were the major component in an ecosystem? We would be stuck to relying on the population level of biological organization to sustain us in the future rather than the ecosystem level. We can thank nature that grasses, at least, are featured at the ecosystem level.

BRIDGING THE GAP BETWEEN "APPLIED" BIOLOGISTS AND CERTAIN PURE SCIENTISTS

What is most needed now is to discover a starting point for bridging the gap between those who work with the species that have already been domesticated for the human purpose, the agriculturists, and those who study nature and natural systems, the biologists and ecologists. Very few of the latter have had any eye at all to the practical applications of their knowledge. There are some important principles from the two synthetic fields of biology, population biology and ecology, that so far have been little applied to agriculture.

Population biology is a synthetic discipline because it is the product, primarily, of three traditions of biology, all of which go back to a synthesis of the work of Mendel and Darwin. It has drawn heavily on the traditions of the biosystematicist, the population ecologist, and the physiological ecologist. This knowledge has been accumulated, and it continues to accumulate mainly for its own sake. Agriculturists cannot take full or immediate advantage of the knowledge accumulated by scientists who have studied the population biology and ecology of wild populations because our domestic species have been too streamlined to meet the demands of human ecology rather than natural ecology. Too many of the ensembles of genes tuned to work well in a natural environment have been stripped away in the process of selection. We should not abandon these "genetic paupers," these domestic crops, in favor of the wild species, but an ecological agriculture is possible only if the natural integrities of wild ecosystems are put to use. If agriculturists could spend more time learning about wild species whose genetic profiles are sufficiently broad to make it on their own in the environment, they may begin to imagine possibilities for the domestic species we use for food and fiber, possibilities which include resistance to insects and pathogens and favorable competition against weeds. Furthermore, we have to think about a sunlight-sponsored fertility. It would be useful if population biologists and ecologists could work together with agriculture in developing one or more domestic ecosystems using certain wild species. It is not an exercise in science fiction to imagine that we will one day take this knowledge and turn to many of our domestic crops and begin to reconstitute their "ecological genes" by crossing them with their wild relatives. In a biological sense we probably have less far to go with the development of new crops from wild species than to reconstitute the old, but there is the cultural reality associated with the traditional foods—the reluctance to change, to adopt a new food. In this respect the cultural barriers may be more powerful than the biological barriers. Because of this, it will be necessary to more completely explore the ecology of agriculture. Such a study would scrutinize the ecological interactions of the various crops and crop rotation patterns in use today. It would be useful to know more of the residual interactions that are a carryover from preagricultural times. Such knowledge could give us an idea of how far we have to go in reconstituting the "ecological genes" that were eliminated long ago in our major crops. These genes will be necessary if we are to develop an ecological agriculture around our traditional species. In the future it will be helpful if we explain whether our research helps us better understand the ecology of agriculture or contributes to an ecological agriculture.

The interface between "pure" population biology/ecology and agriculture has scarcely been explored. If some of the researchers in ecology and in the subdisciplines of population biology—the biosystematicists, population ecologists, and physiological ecologists—were to take the information that has been accumulating for its own sake and relate it to the need to develop an ecological or sustainable agriculture, more progress could be made in the long run than if we restrict ourselves to studying the "ecology of agriculture."

And now we come back to the importance of Darwin. Darwin gave biology an internally consistent unifying concept that was rooted in ecology. Regardless of the fine efforts of countless numbers of agriculturists who have resisted such overemphasis on production, the principle that unites most modern agriculturists has been production, at almost any cost. But an internally consistent unifying concept for the area of applied ecology we call sustainable agriculture does not exist. Once we have said that there is a need to develop a sustainable agriculture, we have made an important leap—we have said that we are no longer willing to discount the future. An agriculture dependent upon high energy inputs, therefore, is no longer consistent with our interests. An agriculture that requires the nearly universal use of chemicals with which our cells have had no evolutionary experience, against which they have evolved no protections, is no longer consistent with our interests. An agriculture that removes people from the land so that the ratio of eyes to acreage is severely reduced is not consistent with our interests either. And most important of all, an agriculture that allows two, three, or four billion tons of U.S. topsoil to run toward the sea each year is unacceptable. The long-term expectations of the land must be considered along with our food needs in our search for an ecologically sound agriculture. The "ecology of agriculture," generally speaking, should be studied in order to diagnose the problems of agriculture rather than to improve the short-run efficiency of food production.

A UNIFYING CONCEPT FOR AGRICULTURE THAT FEATURES INFORMATION OVER ENERGY, SMALL SCALE OVER LARGE

Perhaps the most easily observable physical feature of modern agriculture is the consequence of our denial of the opportunity for the land to experience species succession. To stall succession, the farmer either manipulates or tricks nature. Ground is worked and weed seeds germinate, and then the ground is worked again to destroy the seedlings right before the crop is planted. Many crops are planted at a certain time to avoid the peak of an annual insect infestation. Whether it is management or trickery, the goal is to give an edge, usually to a single favored plant population. All this imposes certain costs, many of which go beyond the field. Loosening the soil every year causes it to erode. When we do not adequately trick the insects or pathogens, we must use more chemicals. Left alone, life would abound. It might not be life that we would regard as having economic importance, but it would abound. The area would feature several stages of plant and animal interaction as it moved toward a dynamic equilibrium recognized as a climax community—tallgrass prairie, oak–hickory forest, or whatever.

The progression I have just described does not happen the way a rock rolls down a mountain slope toward the inevitable rest at the bottom. Biological progression happens under the direction of a molecular information system. What may have begun as bare ground soon becomes a complexity that is scarcely comprehensible. What we see when we walk through prairie is the manifestation of billions of biological bits of interacting information—the molecular DNAs and RNAs of species we will never completely count.

Not only does each individual carry enough instructions to cope in the environment, but also those instructions resonate against the collective background of instructions guiding the other members of the same population. By the time we add in the other populations of a natural community, the total fund of information is awesome. If we were to magnify the molecular "letters" residing in a square mile of tallgrass prairie to the size of print fit for books, we would probably need more space than is present in all the libraries of the world. The biological instructions necessary to produce a corn plant, on the other hand, might fill a large room. After all, one-third of the U.S. corn crop comes from only four inbred lines (Zuber and Darrah, 1980).

Looked at this way, the most obvious difference between crop monoculture and natural polyculture is that the former is information-poor while the latter is information-rich. Furthermore, it is not miscellaneous information, for much of it is keyed to insuring a graceful transition of one species ensemble to the next. When species succession is denied and the juvenile stage is insisted upon each year by the agriculturist, the human–nature split is the most profound. I believe, therefore, that as we begin to maximize information in agriculture, we will eventually feature succession. In such an agriculture a high percentage of the entire genetic program of a "domestic prairie," from birth to death, would be put to use. I expect this emphasis on information will be the starting point of a new paradigm for agriculture that will one day replace the energy-intensive agriculture of today.

Biological information is energy cheap. It has had to depend on sunlight, a dispersed source, and after millions of years of evolution, it has had a chance to evolve toward operating close to thermodynamic limits of efficiency. Compared to the miniaturized and energy efficient information system of biology, our most sophisticated chip-based hardware is large, clumsy, and awkward. In succession agriculture we would be trying to take advantage of the "combination of circuits" that nature has built in over millions of years in "systems" that are self-producing, something we have so far never done. In relying on natural ecosystem integrities, such an agriculture would rely on the DNA language of various species, which operates at an energy cost 10^{21} less than the energy cost for keypunching a comparable "bit" of information on hollerith cards (Delin, unpublished).

Such a saving of energy is not really important. It only illustrates the capability of life forms to become highly efficient in miniaturization. More important are the efficiencies achieved by the products of the code—the parts of the plant that display remarkable efficiencies in structure and function. At the individual level these life forms cooperate with one another, to the benefit of each.

With such an agriculture we would not call ourselves "designers," but "imitators of design," ecosystems that are the culmination of millions of years of trial and error in nature. The distinctly human effort in all this would be to direct these ecosystems toward human use, to encourage yields comparable to those of our traditional crops while avoiding the devastating consequences of monocultures. Such an agriculture would have to be run by naturalists, not industrialists.

Production agriculturists continue to argue for more research money on the grounds that "we must feed the world." Sustainable agriculturists would say instead that "the world must be fed." The surest way to insure that it would not get fed is to make production so expensive that the poor and starving cannot buy the food and the producers cannot afford to give it away. A more sensible strategy would be to develop an agriculture based on the principles of nature's ecosystems and to take advantage of nature's resilience and her efficient use of sunlight once it has been trapped by the chlorophyll molecule. An expensive fossil fuel or nuclear infrastructure should not be necessary to feed people around the world. We should be exporting the principles that emerge from working on the interface between the "pure" sciences of biology and the applied science of agriculture.

I am not completely condemning the researchers who developed production agriculture; sustainable agriculturists will stand on their shoulders and will owe them a debt. But if their successes of the past·50 years or so dazzle us so that we fail to develop a sustainable agriculture, then future generations will become their victims.

Perhaps no one should be blamed for not starting earlier. The kind of agriculture I am advocating had to wait on a level of interaction between ecology and population genetics. It began around 50 years ago. Modern ecology began around 1910, but population genetics became a distinct discipline only in the 1930s and did not mature until the 1960s, for it was not until then that enough research had been done to give us an understanding of the dynamics of the evolutionary process (Stebbins, 1979). This marriage of these two disciplines has now been widely publicized and somewhat more refined. But most agricultural researchers either do not know yet that it exists or they see it as mostly irrelevant. It seems inevitable, however, that practical minds will eventually ponder the implications of such knowledge for agriculture.

The discrepancy between the two economies, the human economy and nature's economy, is real and seems to be widening by the year. Of course, our socioeconomic system controls much of agriculture. As bleak as it now seems, it is worth remembering that the human economic system is full of wild cards. Eighty dollars a barrel oil or the fall of the Saudi Arabian monarchy can change the economic picture, literally overnight. Even if no wild card is ever played, the externalized costs of contemporary agriculture will eventually be more fully understood. When that time comes, it will be important to have a few sustainable alternatives to offer.

I believe that we already know enough to move toward a partial solution of what I have called "the problem of agriculture" (Jackson, 1980). Though some of the basic knowledge has been used to fine tune the traditional crops for increased yield, it has not been used to reduce soil erosion or the use of farm chemicals. It has not decreased the dependency on fossil fuel-based fertilizers or reliance on huge capital investments for farm machinery and equipment. Agriculture will benefit from the inherent virtues of plant mixtures or plant communities—reduced soil loss, resistance to insects and pathogens, appropriate water and nutrient management, and so forth.

This is where an understanding of ecosystems and the dynamics of evolutionary processes are essential. We will be relying on the same tools developed and used by the pioneers in these disciplines, the mathematical and statistical models, the microscopes and greenhouses. We need only change the emphasis.

AN OPTIMISTIC CONCLUSION

A piece of conventional wisdom of our time is that "science and technology are neither good nor evil in their own right; it is how we use them." Even though the statement clearly implies a moral responsibility, we mostly ignore that responsibility and let the momentum carry us into the future. The skeptics who believe, on the other hand, that we have never done anything with science or technology but dig our own graves are so few in number that we somehow feel their opinion does not count, or that their point of view is unworthy of consideration. Let us consider their point of view worthy for a moment and take upon ourselves the task of trying to convince them that to use science and technology to extricate ourselves from the problems in agriculture is worth one more try.

If we were to attempt a thorough assessment of a seemingly benign piece of technology—a wrist watch, for example—long before our assessment is completed, the ecological cost would have mounted, almost astronomically. The modern watch goes back to the Sumerians and Babylonians who divided time into units of 60. Any near-complete technology assessment of a small watch must consider the ecological capital that was spent to provide the "critical mass" of humans necessary for such a time concept to be developed. It must consider the ancient port so silted in that it is now 180 miles inland from where the river emptied in Sumerian times (Stern and Roche, 1974). It must count the salted soils that once supported wheat then salt tolerant barley and finally neither. Although few of us would charge so much against a modern watch, the cost has been real, for most of the options for future generations in that region of the world were closed long ago.

When we first invade a natural ecosystem, we can justify the destruction of a certain amount of biological information because the cultural information that served the human better, in an immediate sense, took its place. But when civilization began to devour culture and also accelerated the reduction in biotic diversity, we eventually began to question the limits.

The Greeks and Romans and the Sumerians and the Babylonians provided us with an abundance of knowledge, the cost of which we will never estimate with any high degree of accuracy. Rather than beat our breasts about the terrible ruin we have wrought, it seems more important for us to take that expensive knowledge accumulated to the present and give that knowledge its proper respect by using it to return stability to the world.

No one gave us the right to destroy the vast wilderness of the North American continent to "build America." Much of the "know-how" in the world that was sponsored at the expense of that wilderness, and other wildernesses besides, is distinctly human, and therefore ours, in a way the wilderness never was. We have always been too much a part of wilderness for it to be ours. Maybe we had to destroy wilderness in order to stand outside far enough to accurately contemplate our origins.

To despise the knowledge we have gained is to regard that knowledge in much the same manner as we regarded the wilderness that made it possible. We should not make the same mistake twice. The Americas, especially, were not really for our taking and plundering in the first place, and no other place will be vouchsafed for us now. The age of ecosystem exploitation must end and, if there is a decent future at all for us, it must be replaced by the age of ecosystem redemption. The ecosystem is a good place to begin to restructure the world. After all, if the split started with agriculture, it is fitting that the healing begin with agriculture.

REFERENCES

Ahrenhoerster, R. and Wilson, T. (1981). *Prairie Restoration for the Beginner*. Prairie Seed Source, Box 83, North Lake, WI, 53064.

Delin, S. Technology or the art of production. (unpublished manuscript).

Jackson, W. (1980). *New Roots for Agriculture*. Friends of the Earth, San Francisco.

Odum, E. P. (1971). *Fundamentals of Ecology*. 3rd ed. W.B. Saunders Co., Philadelphia, p. 52.

Stebbins, L. (1979). Fifty years of plant evolution. In *Topics in Plant Population Biology*. Otto T. Solbrig et al. (eds)., Columbia University Press, New York.

Stern, K. and Roche, R. (1974). *Genetics of Forest Ecosystems*. Springer-Verlag, New York.

Weaver, J. E. (1954). *North American Prairie*. Johnsen Publishing Co., Lincoln, Nebraska, pp. 210–211.

Wittwer, S. (1981). The 20 Crops that protect the world from starvation, *Farm Chemicals*.

Zuber, M. S. and Darrah, L. L. (1980). *Ann. Corn Sorghum Res. Conf. Proc.* **35:** 234–249.

Index

Aboveground net production, *see* ANP
Actinomycetes, 88
Agricultural Products, 8
  food, 8
  market commodities, 8
Agricultural Systems, 122, 179–185
  animal powered, 122
  classification of, 184
  commercial, 184
  computer experimentation, 180
  human powered, 122
  models of, 180
  reasons for study, 179
    improvement, 179, 182
    operation, 179–180
    repair, 179–181
  tractor powered, 122
Agriculture, 14, 15, 18, 19, 29, 121–130,
    172–176
  commercial, 14
  compared to natural systems, 174
  complexity, 174
  development of in the Midwest, 7
  drain on life support, 6
  human factors, 174
  impact on Linsley Pond, 8
  industrial, 5
  management, 174
  modern, 74
  organic, 29
  potential of modeling in, 170
  practice of, 172
  preindustrial, 5
  role of economics, 14, 15, 18, 19
  significance of statistics, 19
  village level, 14, 15
Agroecosystem management, 59, 60, 65,
    75–78
  animal grazing systems, 65

chemical control, 78
conventional tillage, 79
interplanting, 77
management intensity, 59
modified management, 76
negative impact, 75
no-tillage, 79
positive interactions, 76
residue management, 79
resource inputs, 59, 60
tillage practices, 79
Agroecosystems, 5–14, 15, 28, 43, 74, 84,
    105–118, 166, 167, 187–206, 214
artificial ecosystems, 74
chemical and mechanical disturbances,
    79
contribution to life support, 6, 8
defined, 14, 15
of developed countries, 14
domesticated ecosystems, 5
farms system, 74
fuel and fertilizer inputs, 84
  minimized, 84
functional processes, 85
journals, 166
in less developed countries, 14
long-term fertility, 11
manner of control, 9
mechanized, 187, 191, 201
of monocrops, 15
net primary production, 204
power density level of, 5
primary production, 74
properties of, 5
rice production systems, 204
role of economic factors, 14, 15
secondary production, 74
stability, 11
stress, 190

Agroecosystems (*Continued*)
  substructure of, 175
  tropics, 201
  village level, 14, 15, 28
Agronomic Effects of Erosion, 135–139
  and crop yield reductions, 135–137
    masked by management, 136
    related to soil profile, 136–137
  ecosystem effects, 140–142
  effects on water, tilth, and fertility, 137–139
    organic matter, 138–139
    poor soil fertility, 139
    soil water storage, 137–138
Air Pollution, 197, 198, 206
  losses from, 198
  oxidants, 197, 198, 206
  sulfur dioxide, 197, 198, 206
Akron, Co. Tillage Plots, 100, 101
  nitrogen accumulation, 100, 101
    ammonium, 100, 101
    bulk density, 100, 101
    concentration, 100, 101
    nitrate, 100, 101
Allelopathic Chemicals, 76–77
Allocation-selection, 206. See also A-selection
Alternative Tillage, 146, 152. See also Minimum Tillage; No-Till
  effects, 152
    compaction, 152
    residues, 152
    soil temperatures, 152
  mimic natural systems, 152
  nutrient balance, 146
  persistence, 146
  productivity, 146
Animal Confinement Systems, 64
  animal husbandry, 64
Animal Grazing Systems, 65
  herbivores, 65
  species composition, 65
ANP:
  Abies, 32
  alfalfa, 28
  angiosperms, 32
  banana, 33
  barley, 29
  beans, 29
  bermuda grass, 28

betel garden, 41, 42
Betula, 32
C3 forage grasses, 27, 28
C3 marshes, 24
C4 forage grasses, 27, 28
climate, 46
conifers, 32
crop, 22
Cynodon dactylon, 28
Cyperus papyrus, 23, 24
defined, 15–17
ecological measures, 14
  relative to economic yields, 14
Erythrina, 42
extrapolation to NPP, 16, 17
Fagus, 32
floodplains, 20
forage crops, 27
forests, 20
grassland, 20
grassland biome, 20, 27, 28
hay, 27, 28
intercrops, 35–39
lawns, 28
Lolium perenne, 28
maize, 21, 29
manioc, 33
marshes, 20, 23, 24, 27
Medicago, 28
millet, 29
mixed cropping, 32, 33
Moringa, 42
napiergrass, 27, 28, 45
natural community, 14
oats, 29
oil palm, 33
Pennesetum purpureum, 28, 45
Picea, 32
Pinus, 32
Populus, 32
Pseudostuga, 32
rice, 23, 24, 25
rye, 29
ryegrass, 28
Sesbania, 42
sorghum, 29
soybean, 29
Spartina alterniflora, 23, 24
subsidized, 20
sugar cane, 33

sunflower, 29
temperate marsh, 24
trees, 32
tropical crops, 32
Typha latifolia, 23, 24, 25
unsubsidized, 29
upland seed crops, 27
upland temperate forests, 32
variability, 20, 21
wheat, 18, 29, 31
Zizania aquatica, 23, 24, 25
Zizaniopsis maliacea, 23, 24
see also NNP; Production
A-selection, 189, 198, 200
Assimilation, 17

Babylonians, 220
Bacteria, 88
Bananas, 213
Barley, 220
Beans, 111, 112, 115
Below ground production, 13
Beneficial Insects, 78
Biological Pest Control, 191, 206, 116–
    117. See also Integrated Pest
    Management
Biology, modern, 211
Biosystematics, 215, 216
Bristle cone pine, 84
    delta 13C, 84
    ligno-cellulose, 84

Carbon:nitrogen ratios, 154
Cash crop, 9
    long term productivity, 9
Cassava, 115, 213
Chemical Pesticides, 61
Climate, 17
    effect on ANP, 17
Coconuts, 213
Coevolution, 78
    insect-plant, 78
    major theme insect ecology, 78
Colonization and Succession, 97
    microbial and faunal, 97
    random process, 97
    synthetic logs, 97
        arthropods, 97
        establishment, 97, 98
        invasion, 97, 98

Community, 13, 14, 16, 21, 41–43, 45,
    47, 217, 219
    betel gardens, 41, 42
    climax, 217
    crop, 13, 43
    intercrops, 35–39
    measuring growth, 16
    mixed cropping, 21
    monocrop, 21
    natural, 13, 21
    plant, 219
    structure, 14, 21, 42
    swidden agriculture, 40, 41
Compaction, 102. See also Tillage Practices
Comparison of Natural and Cultivated
        Ecosystems, 149–151
    abiotic factors, 149, 150
    aiotic factors, 150, 151
Competition, 77
Conservation Farming Practices, 8, 9
    general effect, 9
Conservation Tillage, 201, 202
    minimum tillage, 201
    no tillage, 201, 202
    see also Minimum Tillage; No-Tillage
Consumers, 55–68, 75, 77–78
    coarse grained, 59, 60, 61, 65
    fine grained, 60, 65
    herbivore optimization process, 66
Contour Plowing, 197
Control of consumer pests, 61
    chemical pesticides, 61
    corn rootworm, 61
    plant resistance, 61
    potato insecticides, 61
Conventional Tillage Farming, 9, 10
    comparison to no-tillage, 9, 10
        ecosystem level processes, 9
        yields, 9
Corn, 213, 214, 217
Cost, 219, 220
    ecological, 219
    external, 220
Creation, 211
Crop Drying, 129, 130
    fossil energy, 130
    kcal per Kg, 129, 130
    solar energy, 130
        advantages of, 130
Crop Growth Rate, see Growth Rate

Crop Physiology, *see* Physiology
Cropping Systems, 84, 107, 109–112, 117, 118
Crop Residues, 152
Crop Rotations, 197
Crop Yield Variation, 198–200
  yield instability, 200
Cultivar:
  distribution of, 14
  phenotypes, 14
  see also specific crops under ANP;
    Growth Rate
Cultivated Ecosystems, 146
Cultivation, 75
Cultural Control, 202
Cultural Practices, 128, 129
  conventional tillage, 128, 129
  energy use, 128, 129
  erosion, 129
  moisture, 129
  no-tillage, 128, 129
  organic matter, 129
  pests, 128, 129
    insects, 128
    pesticides, 128
    slugs, 128
    weeds, 128
Cybernetics of Ecosystems, 5, 116
Cybernetic Systems, 10
  basic features, 10

Darwin, Charles, 211, 212, 215, 216
Decay Curve, 86
Decisions, 106–114, 117, 118
  control, 106–110, 112–115, 117, 118
  design, 106–111, 113, 114, 117, 118
  determinants, 106, 110, 111, 113, 115
Decomposers, 89
Decomposition, 75, 79–80, 147, 152, 154
Decomposition Phases, 97
  colonization and succession, 97
    random element, 97
    synthetic logs, 97
  exploration, 97
  invasion, 97
  post invasion, 97
Decomposition Processes, 84, 90, 92
  microorganisms and, 93
    secondary metabolites, 93

negative exponential model, 89, 90, 92
  rate factor (k), 90
Determinants, 105–118
  defined, 106
  types, 106, 109, 110, 112, 115
Diversity, 5, 187, 188, 214, 220
  biotic, 214, 220
  intraspecific, 188
Dryland Cropping, 86

Ecological Capital, 220
Ecological Factors, 14
Ecological Niche, 187
Ecological Principles, 153–155
Ecological Succession, 187
Ecology, 7, 212, 215, 216, 219
  holistic systems approach, 7
  interdisciplinary, 7
Economics, 219
Economic yield, 13
Ecosystem analysis, 75
  allelochemical regulation, 75
  energy flow, 75
  nutrient cycling, 75
  response to perturbation, 75
  species richness, 75
Ecosystem Degradation, 149
Ecosystem Effects of Erosion, 140–142
  field level, 140–141
    cation exchange capacity, 140
    nutrient loss, 140
    organic matter losses, 140
  landscape level, 141
    increased inputs, 141
    land conversion, 141
  watershed level, 141
    channel degradation, 141
    sediment deposition, 141
Ecosystem Persistence, 146
Ecosystems, 5–10, 15, 43, 73, 121, 130, 158, 179, 180, 213, 214, 217, 218, 220, 221
  and agroecosystems, 15
  control, 158
  cybernetic nature of, 5, 10
  domesticated, 5, 8
  fabricated, 5, 8
  functional properties of, 74
    energy storage and utilization, 74

nutrient conservation, 74
  regulation of biotic diversity, 74
holistic approach, 74
models, 168
  completeness of, 168
natural, 5, 179
substructure of, 175
village, 28
Efficiency of Systems, 182
Energy, 5, 8, 105, 107, 110, 113, 115,
        118, 121–130, 187, 191, 193, 197,
        200, 203, 206, 218
auxiliary, 5
biomass, 123
efficiency, 8
flow, 105, 116, 124–125
fossil, 122, 204
heat, 122
human, 122
infrastructure, 218
inputs, 191
organic farming, 203
relative to farm production, 191
sources, 5
wood, 122
Energy Budget, 16, 43
defined, 16
relative to field data, 16
relation to production, 16
Energy Efficiency, 123, 204
corn, 123
natural vegetation, 123
potatoes, 123
wheat, 123
Equilibrium, 217
Erosion, 147, 151, 153
Erosion Reduction, 102. *See also* No-Till;
        Tillage Practices
Essential Nutrients, 147
  uptake and circulation, 147
Eutrophication, 8
Evaporation, 153, 154
Evolution, 211

Farm machinery, 219
Farm Size, 191, 193
  and energy, 191, 193
  and fertilizers, 193
  intermediate size, 193

Farm System, 105, 107, 111, 113, 114
Fertilizers, 75, 127, 189, 191, 193, 195,
        196, 197, 202, 203, 204, 206, 219
effectiveness, 195, 196
micronutrients, 201
nitrogen, 127, 196, 202
phosphorus, 127
potassium, 127
use, 193, 195, 196, 197, 206
Fossil Fuel Combustion, 84
atmospheric carbon dioxide, 84
Fungi, 88
hyphae, 90
  transport of nitrogen, 90

Genesis, 211
Global Population, 214
Grasses, 212–214
Greeks, 220
Green Revolution, 204
Growth Rate:
alfalfa, 28
averages, 22, 30, 44
banana, 33
barley, 29
beans, 29
bermuda grass, 28
betel garden, 41, 42
C3 forage grasses, 28
C3 marshes, 24
C4 forage grasses, 28
crop, 22
Cynodon dactylon, 28
Cyperus papyrus, 24
defined, 16
Erythrina, 42
factors affecting, 24–31
grassland biome, 20, 27, 28
hay, 28
intercrops, 35–39
Lolium perenne, 28
maize, 29
manioc, 33
marshes, 24
maximum, 22, 27, 45
measurements, 22
Medicago, 28
millet, 29
Moringa, 42

Growth Rate (*Continued*)
  natural communities, 14
  napiergrass, 28
  oats, 29
  oil palm, 33
  Pennesetum purpureum, 28
  in physiological time, 16
  rice, 24, 25
  rye, 29
  ryegrass, 28
  Sesbania, 42
  sorghum, 29
  soybean, 29
  Spartina alterniflora, 24
  subsidized, 41
  sugar cane, 33
  sunflower, 29
  temperate marshes, 24
  Typha latifolia, 24
  wheat, 29
  yields, 31
  Zizania aquatica, 24
  Zizaniopsis maliacea, 24
  see also Production

Harvest Index:
  barley, 29
  defined, 17
  maize, 29
  millet, 29
  oats, 29
  rye, 29
  sorghum, 29
  soybean, 29
  sunflower, 29
  wheat, 29
Herbicides, 75
Herbivores, 148
Heredity, 212
Heteropolycondensates, 96
Heterotrophs, 88
  actinomycetes, 88
  bacteria, 88
  fungi, 88
Hierarchical levels, 96
  agroecosystem dynamics, 96
  simulation models, 96
Hierarchy, 57, 66
H. I., see Harvest Index
Homeostasis, 146

Human-nature, 217, 221
Humified substances, 97
  humads, 97
Hutton, James, 211
Hypereutrophication, 8

Ideotypes, 23, 24, 31, 43
  leaf area index, 25, 31
  rice, 25
  selection, 23
  wheat, 31
Immigration of propagules, 151
Infiltration of water, 149, 153
Information, 105, 107, 109, 113–118, 217, 220
  behavioral, 116
  biological, 220
  cultural, 220
  flow, 107
  genetic, 116
  processes, 105, 107, 109, 113–118
  systems, 217
  theory, 113, 116
Input and Output Environments, 5, 7
Insect Herbivores, 77
  compensatory growth, 79
  crop consumption, 79
  economic importance, 77
  economic thresholds, 78
  low-level grazing, 78
  plant-insect mutualism, 79
Insecticides, 75
Integrated Pest Management, 62, 78, 202
  cultural control, 202
  biological control, 202
  integrated control, 202
  orchard crops, 62
  row crops, 62
Intercropping, 111, 112. See also Mixed
    Crops
Irrigation, 130
  drip, 130
  energy and, 130
  sprinkle, 130
  water, 130
    limiting factor in crop production, 130
    conserving techniques, 130
Isaiah, 213

Juhm, see Swidden Agriculture

Labile Minerals, 148
LAI, see Leaf Area Index
Land Equivalent Ratio, 34
Land Resource Base, 134, 135
  humid East, 135
  United States, 134
Landscape Pattern, 55–68
  design, 56
  dynamic, 60
  heterogeneous, 60
  remote sensing, 56
  spatial heterogeneity, 56
  static, 60
  structure, 56
Leaching, 151, 154
Leaf Area Index, 21, 26, 31, 41
  defined, 21
  swidden, 41
  wheat, 31
Legumes, 128, 212–214
  green manure, 128
  sweet clover, 128
  winter vetch, 128
Life Support Systems, 6, 8
  spaceship earth, 8
  drain by agriculture, 8
Limited Till Farming, 9
Limiting resource, 78
Litter, 153
Livestock, 107, 109, 112, 114
Lyell, Charles, 211

Maize, 109, 111, 112, 115
Malthus, Thomas (Parson), 211
Management-Landscape Continuum, 60
Management regimes, 85, 87
Manure, 127
  equivalency to inorganic fertilizer, 127
  as fertilizer, 127, 128
  green, 128
  livestock, 127
  nitrogen, 127
  phosphorus, 127
  potassium, 127
  transportation, 127
Marsh, 213, 214. See also ANP, Growth
    Rate
Mendel, Gregor, 212, 215
Metabolic Activity, 88
  faunal, 88

  microbial, 88
Microarthropods, 100
  eudaphic collembola, 100
  predatory mites, 100
  see also Tillage Practices
Microbial Activity, 85
Microbial Decomposers and Mineralizers,
  147
Microbial-Faunal Interactions, 97
  spatial arrangement of organisms, 97
Microbial Production and Turnover, 92–95
  carbon:nitrogen ratio, 92
  models of, 93–95
  N-mineralization, 92
    gaseous losses, 92
    oxidation and reduction, 92
Microbial Populations, 102
  management, 102
  nutrient turnover, 102
Mineralization, 88, 152, 154
Minimum Tillage, 153
  element uptake, 153
  microbial activity, 153
  plant growth, 153
Mixed Cropping, 200, 201
  alternate row cropping, 201
  interseeding, 201
  strip cropping, 201
Mixed Crops, 47
  ANP, 33
  bean-maize, 37
  betel vine, 41
  Cajanus cajan, 38
  caster bean, 39
  consistency of yields, 36
  cowpea, 38
  dry land, 40
  efficiency, 33
  Erythrina, 42
  field bean, 39
  groundnut, 36
  Hybiscus cannabis, 38
  Lablab purpurea, 39
  land equivalent ratio, 34
  maize-soybean, 35
  millet, 36
  Moringa, 42
  mung bean, 38
  overyield, 34
  relative labor, 35

Mixed Crops (*Continued*)
  relative yield, 34, 35, 36
  red gram, 37–40
  reliability, 36
  Rincinus communis, 39
  Sesbania, 42
  Sorghum vulgare, 38
  soybean, 35
  swidden, 40
  Vigna radiate, 38
  V. trilobata, 38
  V. unquiculate, 38
  yields, 33, 36, 41
Modeling, 157, 169, 170, 172
  need for interdisciplinary research in,
      170, 172
  potential of, 170, 172
  as unifying procedure, 169, 170, 172
Models, 109, 110, 114, 159, 179–186
  of agricultural ecosystems, 184–185
  building and testing, 182
  deficiencies of, 166
  of natural ecosystems, 184–185
  nature of, 160
    simplifications, 160
    nonuniqueness, 160
    volatility, 160
  optimization, 114
  purposes of, 161–163
    exploration, 161–162
    explanation, 162
    projection, 162, 163
    prediction, 162, 163
  simulation, 114
  validation, 182–183
  verification, 183
Models of Microbial Production and
      Turnover, 93–95
  PHOENIX model, 95
  soil organic matter (SOM) model, 93
Moisture Infiltration and Storage, 87
Monocrops, 13, 15, 35, 47
  vs. intercrops, 35, 37
Monoculture, 76, 213, 214, 217, 218
Multiple Use Systems, 66
  management goal, 66
  size, 66
  see also National Forests
Mutualistic Associations, 77
  mutualism, 77
  mutual tolerance, 77

Mycorrhizal Fungi, 79
  plant-fungus mutualisms, 79
  spore dissemination by arthropods, 79
Mycorrhizal Networks, 10
  improve nutrient retention, 10

National Forests, 66
  quasi-natural systems, 66
Natural Selection, 211
Net Primary Production, *see* ANP
Nonintensive Management Levels, 64
Nonrenewable Resources, 209
No-Tillage, 9, 92, 128, 129
  comparison to conventional tillage, 9,
      128, 129
  ecosystem level processes, 9
  energy use, 128, 129
  negative exponential model of
      decomposition, 92
  pests, 128, 129
  soil erosion, 129
  soil moisture, 129
  soil organic matter, 129
  yield, 9
Nontilled Ecosystems, 146
NPP, 16, 18, 44
  correlation with climate, 18
  defined, 15–17
  relation to ANP, 16
Nutrient Cycle Model, 75
Nutrient Cycles, 105, 116
Nutrients, 19, 20, 27, 29, 42, 46, 84, 126–
      128
  effect on ANP, 19
  nitrogen, 84
  phosphorous, 84
  soil, 126–128
  subsidies, 20, 27, 29, 42, 46
  sulfur, 84
  see also Fertilizers

Odum, Eugene, 73, 74
  fundamentals of ecology, 73
  holistic approach, 74
On the Origin of Species, 210, 212
Organic Carbon, 84
  global flows, 84
  losses of, 84
    fossil fuel combustion, 84
    tropical deforestation, 84
  pools, 84

Organic Farming, 9, 202–203
  decline in surpluses, 9
  domestic effects, 9
  economic effects, 9
  energy, 203
  nitrogen fertilizers, 203
  organic practices, 202, 203
  widespread adoption, 9
Organic Nitrogen, 84

Paradigms, 211, 218
  "instant creation," 211
  new agricultural, 218
Pastoralism, 187
Pastoral system, 105, 113, 114
Pesticides, 201, 202, 204
  herbicides, 75, 202, 204
  insecticides, 201, 204
Pest Management Consultants, 201
Pests, 121, 122
  birds, 121
  breeding resistance, 129
  control of, 129
  insect, 121, 122
  insecticides, 129
  mammals, 122
  pesticides, 129
  plant pathogens, 121, 122
  weeds, 122
Phaseolus coccineous, 112
Phaseolus vulgaris, 112
Phosphorus Model, 95
  cycling, 95
  excreta, 95
  inorganic, 95
    stable, 95
    labile, 95
  organic, 95
  soil micro-, meso-, macrofauna, 95
Photorespiration, 200
Photosynthesis, 31, 45, 46, 200, 206
  C3 *vs.* C4, 45, 200
  photorespiratory losses, 200
  rates, 31
Physiological Ecology, 215, 216
Physiology, 16, 43, 46
  crop, 43, 46
  processes, 16, 43
  time units, 16, 46
Plant Canopies, 147

Plant Genetics, 200
  cell culture, 200
  protoplast fusion, 200
  recombinant DNA, 200
Plant Residues, 154
  carbon:nitrogen ratio, 154
Plowing, 75
Polyculture, 201, 206, 213, 217
  multigrade, 201
  see also Intercropping
Population Biology, 215, 216
Population Ecology, 215, 216
Population Genetics, 219
Positive Interactions, 75
  biotic persistence, 75
  cooperative relationships, 75
  maintenance of functional integrity, 75
  self-regulation, 75
Potatoes, 213
Power, 124
  hand, 124
  horse, 124, 125
  machinery, 124
  tractor, 124, 125
Prairie, 213, 214, 217
Prairie States, 213
Predator-Prey, 116
Predators, 148
Primary producers, 75–77
Processes, 10, 57
  ecosystem level, 10
  self organizing and maintaining, 10
Producers, 55–68
Production, 13, 15, 16, 19, 43, 44, 46
  annual variation, 20, 47
  ANP, 13, 15, 16
  assimilation, 17
  comparisons, 23
  economic yields, 15
  extrapolation to NPP, 16
  gross, 15
  NPP, 15
  nutrients, 19
  respiration, 17
  roots, 16
  swidden agriculture, 40
  time units, 16, 43, 46
Recycling of nutrients, 155
Residue Management, 197
Residues, 84
  decomposition, 86

Residues (*Continued*)
  leaf, 86
  root, 86
  stem, 86
Resistance to Insects and Pathogens, 215, 219
Rhizosphere, 97
  biota, 97
  fungal symbionts (mycorrhizae), 97
  root cortex, 97
  roothairs, 97
  roots, 97
Rice, 214
Romans, 220
Rooting Structure, 155
Root Systems, 151
Row Crops, 61
  corn rootworm, 61
  plant resistance, 61
  potato, 61
Runoff, 149

Secondary Pest Outbreaks, 202
  and pesticides, 202
Sedimentation, 153
Selection, 5, 189
  artificial, 5
  A-selected, 189
  K-selected, 189
  natural, 5
  R-selected, 189
Shifting Cultivation, 187
Slash and Burn, *see* Swidden Agriculture
Socioeconomic System, 106, 107, 110
Soil, 79
  determinant of primary production, 79
  nutrients, 79
  nutrient storage reservoir, 79
  organic matter, 79
  structure, 79
Soil Conservation, 197
  and tariff on exports, 202
Soil Erosion, 129, 134–137, 141, 142, 195, 196, 197, 202, 219
  control of, 197
  future of erosion, 142
  legumes, 137, 142
  rates, 196
  rill, 196

sheet, 196
T values, 135, 141
water, 195, 202
wind, 195, 202
world food and fiber demands, 134
Soil Fertility, 197
Soil Management Practice Tax, 202
Soil Organic Matter, 83, 84, 87, 89, 129, 147
  depletion, 84
  income, 89
  labile, 89
  loss, 89
  nonlabile, 89
  stabilizing by clay surfaces, 87
  storage of carbon, 84
Soil Structure Protection, 102. *See also* Tillage Practices
Soil Temperature, 152, 154
Soil Water Loss, 129, 149
Sorghum, 111, 112
Southern Pine Beetle, 63. *See also* Tree Plantations
Species Diversity, 151
Stabilized Minerals, 148
Subsidies of Energy and Material, 74
Substrates, 89, 96
  primary, 96
  secondary, 96
Subsystem, 158, 168
  component interactions, 169
  connections, 168
    direct, 168
    indirect, 168
  control, 158
  management, 168
Succession, 117, 217
Sugarbeets, 213
Sumerians, 220
Summer Fallow, 86
Supersystems, 167
  agricultural, 167
  composition of, 167
Sustainable Agriculture, 145
Sweet Potatoes, 213
Swidden Agriculture, 40
  fallow periods, 40, 41
Synchrony of Plant and Microbial Activity, 150, 154
System Improvement, 181–182

Systems, 158, 166, 167
  agricultural, 167
    classification of, 167
  connections, 168
    direct, 168
    indirect, 168
  control of, 158
  management in relation to subsystem,
    168
  need for definition of, 166
  substructure of, 175
Systems Analysis, 157
Systems Approach, 163–165, 174–175
  strengths, 174–175
  systems analysis, 164
    nature of, 164
    opinions of, 164, 165
  systems theory, 165
    in agriculture, 165
  weaknesses, 174–175
Systems Principles, 181

Technology and Science, 220
Terraces, 197
Tillage Practices, 99
  conventional tillage, 99
  minimal tillage, 99
    stubble mulch, 99
    zero tillage, 99
  no-tillage *vs.* conventional tillage, 99
    compaction, 102
    earthworms, 100
    erosion, 102
    microarthropods, 100
    nitrifiers, 99
    soil structure, 102
Translocation, 150
Tree Plantations, 62, 63
  loblolly pine, 63
  silviculture, 62

  slash pine, 63
  woody biomass plantations, 63
Tropics, 106, 109, 111, 112, 118
  Central America, 106, 111, 112, 118
  El Salvador, 112
  Honduras, 111, 112

Urbanization, 8
  impact on Linsley Pond, 8

Vegetation Canopies, 153
Villages, 14, 15
  as ecosystems, 15
  function of, 15, 28
  mixed cropping, 36
Vigna spp., 112

Weeds, 77, 151
  and beneficial insect populations, 77
  beneficial weeds, 77
  and erosion, 77
  nutrient reservoir, 77
  and pest problems, 77
Wheat, 214, 220. *See also* ANP; Growth
    rate
Wilderness Systems, 67, 220

Yields, 8, 13, 18, 24, 27, 28, 29, 30, 31,
    33, 34, 47
  economic, 13, 47
  evolution of, 24, 27, 30, 47
  long-term sustainable production, 8
  mixed cropping, 33, 34
  and photosynthetic rate, 31
  relative, 29, 34
  rice, 28
  short term production, 8
  wheat, 18, 30

Zero-tillage, 84. *See also* No-Tillage; Tillage
    Practices